PRINCIPLES OF X-RAY METALLURGY

Principles of X-Ray Metallurgy

T. KOVACS.

Senior Lecturer in Materials Science and Technology,
Hatfield College of Technology

SPRINGER SCIENCE+BUSINESS MEDIA, LLC

Library of Congress Catalog Card Number: 73-81852

ISBN 978-1-4899-5572-2   ISBN 978-1-4899-5570-8 (eBook)
DOI 10.1007/978-1-4899-5570-8

Filmset by Photoprint Plates Ltd
Wickford, Essex.
Printed in England by
J. W. Arrowsmith Ltd, Bristol

Contents

Preface

During the past thirty years, the use of X-ray diffraction photography has gained wide acceptance both in production processes and in applied research. One of the obstacles to an even wider industrial application is that many undergraduates are not as familiar with the subject as might be hoped. By presenting this book, the author hopes to dispel the common misconception that X-ray diffraction techniques are best taught in postgraduate courses.

The author's treatment of the whole subject of X-ray metallurgy is intended as a useful guide to Degree and National Diploma students. All the basic principles are explained and illustrated by practical examples, whereas some rigorous theoretical derivations and very specialised procedures have deliberately been omitted in the interests of simplicity. For instance, since the book is designed for students who are unlikely to be engaged on structural work early in their industrial careers, descriptions of structure analysis methods have not been included.

Instead of the usual references given within the text or at the end of each chapter, the reader is referred to the Bibliography, which provides lists of both general and specialised books.

The author is greatly indebted to Dr D. Lewis of the University of Surrey for his constant encouragement, and to Mr G. Smith, Head of the Department of Industrial Engineering at Hatfield College of Technology, for the help he gave during the preparation of this book. In addition, the author wishes to express his gratitude to Mrs Joan Frost and Mrs Barbara Cohen, who so painstakingly typed and retyped the manuscript.

HATFIELD, 1968   T.K.

1

Introduction to fundamental crystallography

1.1 THE DEVELOPMENT OF MODERN CRYSTALLOGRAPHY

The name crystal has its origin in the Greek word *krystallos*, meaning ice. In the early Greek civilisation it was thought that rock crystal, to which the name *krystallos* first referred, was made of such intensely frozen ice that it could never melt again.

Ice had a fascination for the great astronomer Johannes Kepler. During his stay at the court of Emperor Rudolph II in 1611, Kepler published a small pamphlet: *A New Year's Present; On Hexagonal Snow*. The appearance of regularly shaped snow crystals suggested to Kepler that external regularity of shape might be due to the internal orderliness of some extremely small, equal, brick-like units. The great astronomer probably realised that his hypothesis could lead to an atomic theory — which, however, he distrusted for its lack of empirical foundation. Nevertheless, Kepler was led to the idea of close-packed spheres as the basic building units of crystals. The first pictures of space lattices can be found in his work.

Owing to the complete lack of atomic theory at the time, progress in crystallography was restricted to investigations of external shapes of crystals, without undue concern about their internal arrangements. Guglielmini made observations on the constancy of cleavage directions in calcite, and in 1669 Niels Stensen discovered the existence of characteristic angles between crystal faces.

The work of Rene Just Haüy is, however, often regarded as the foundation of systematic crystallography. Haüy recognised that the rich variety of external appearance of crystalline matter might possibly be based on a common internal structure; in other words, the crystal's shape is really the external expression of the internal structure of the substance.

In his publication (1784) entitled *An Essay about a Theory on the Structure of Crystals Applied to Several Types of Crystallised Substances*, he introduced the principle of constancy of cleavage. Finding that a cleavage nucleus can be extracted from many crystalline substances, he put forward the principle that continued cleavage must ultimately lead to the smallest unit; and these smallest units, when repeatedly joined together (not unlike bricks), may form the overall external shape of the crystal. A structure of this kind involves a space lattice, and the next step was the deduction of the laws governing the geometry of crystal faces, which so far had only been studied empirically.

Haüy's discovery of the basic elemental shape is directly related to our present-day knowledge of unit cells and their implications in crystals. However, it cannot be regarded as an indication or prediction of atomic structure, for this did not come about until the year 1824. Ludwig August Seeber then introduced the idea that crystalline substances are composed of chemical atoms or molecules whose interatomic distances are determined by a delicate balance of attractive and repulsive forces, thus forming a system of equilibrium. Seeber's work was far ahead of its time: the kinetic gas theory, regarded as the beginning of atomic theory in physics, appeared about 30 years after Seeber's publication. The geometrical implications of Seeber's work were taken up readily by Frankenheim and by Auguste Bravais. In 1850 Bravais developed the 14 pure translation lattices, and by 1890 Fedorov and Schoenflies had completed the derivation of all possible space groups, as are used today in structural investigations.

On 8th November 1895, Wilhelm Conrad Röntgen connected the terminals of an induction coil to the electrodes of a Crookes tube in an experiment to study high-voltage discharge phenomena. But this time, the normally routine experiment took a very curious turn indeed. In the darkened room, a screen of barium platinocyanide placed some distance away from the tube displayed an unexpected fluorescent glow.

The following six weeks of intensive research proved that there were some invisible rays passing through the air causing the screen to emit visible light. Because of their unknown nature, they were called *X-rays*. In less than a month, X-rays were generated and

2

studied all over the world. The first applications of X-rays were in the field of medical radiography: an X-ray generating tube was placed on one side of an object and a photographic plate on the other, and the differential absorption of the rays caused a shadow picture to appear on the photographic plate. The medical world quickly accepted the newly discovered rays for radiographic purposes, without precisely understanding the nature of the radiation.

A turning point came in 1912, when the mystery nature of X-rays was first explained in real scientific terms. Until that date, some very eminent physicists, including H. W. Bragg, had held firmly to the corpuscular theory of X-rays, while others had believed in the wave nature of this, as yet, inexplicable radiation. Experiments on ionisation effects of X-rays, on the one hand, supported the protagonists of the corpuscular theory; whereas experiments on polarisation and on photo-electric effects confirmed the electromagnetic behaviour of X-rays, to the immense satisfaction of the opposing side.

Naturally, the contradictions of the two theories are now fully reconciled by the quantum theory.

In the Spring of 1912, von Laue was investigating the undulatory nature of X-rays and reached the conclusion that, if the wavelength values calculated by Sommerfield were correct (4×10^{-9} cm), then Seeber's theoretical crystals would diffract X-rays in a manner similar to the diffraction of a light beam by optical gratings.

The discovery of X-rays by Röntgen revolutionised the medical science of the day; Laue's discovery of diffraction of X-rays by crystals transformed crystallography from its contemporary descriptive state into a forward-looking analytical science. Laue's results proved beyond doubt the wave nature of X-rays, and also provided evidence of the atomic structure of crystalline matter.

In the meantime, experiments by Barkla, Bragg, Mosely and Darwin on reflected X-ray beams brought out a new feature — the characteristic line spectrum. The result of this discovery made structural analysis considerably easier, as the difficulty in relating indices to the 'Laue spots' (the photographic results of diffracted X-rays beams) could be overcome by the choice of radiation with a known wavelength.

Until quite recently, X-rays were the most important means of investigating unknown crystal structures, chemical compositions and polycrystalline orientations: they became the most important single tool of solid state physics.

3

1.2 A SURVEY OF THE CRYSTALLINE STATE

For the most part, solids are crystalline. This means that, in solid substances, atoms, ions or molecules are arranged in an orderly manner, which is repetitive in three directions. If the repetitive pattern appears in one or two directions only, the substance is said to be semicrystalline.

Crystalline solids often occur as polycrystalline aggregates, containing large numbers of crystals of various sizes, often microscopic, orientated in a random or in a preferential manner. It has been found that many monocrystals, particularly metallic ones, are composed of subcrystallites of sizes between 10^{-4} cm and 10^{-5} cm. Within the subcrystallites the orderly arrangements of atoms are very near perfect, but the subcrystallites themselves are somewhat disorientated relative to each other by an angle of a few minutes. The crystalline state, being the lowest energy condition, is the natural state into which atoms, ions, or molecules arrange themselves at temperatures below their melting point.

The study of the internal symmetries of atomic arrangements leads to an understanding of the symmetries, or asymmetries, of physical properties of crystals. Some crystalline substances have sets of physical properties which are direction-dependent, although the atoms composing the substance are themselves isotropic. This experimental anisotropy is the direct result of the internal order of atomic arrangements.

The external symmetry of a crystal may indicate the internal arrangement of the atoms. A plate-like crystal is likely to contain long molecules in a fibrous arrangement, whilst spherical molecules or isotropic ions are usually found in equidimensional crystals.

There are many thousands of substances whose structures are known in great detail. When these crystals were grouped according to common physical properties, it was found that they could be classified by the types of forces which act between the atoms occupying pre-ordained positions in space.

1.2.1 IONIC CRYSTAL BOND

It is assumed that ions are electrically charged atoms (achieved by either the subtraction or addition of a number of electrons) with spherical symmetry; and that they attract or repel each other, according to simple electromagnetic laws.

4

The interaction between two ions x and y may be represented by two expressions. The primary interaction is the electrostatic force f between the ions, given by

$$f = \frac{e_x e_y}{r_{xy}} \tag{1.1}$$

where e_x and e_y represent the charges, and r_{xy} is the distance between the centres of the ions. The corresponding potential energy ϕ for each of the two ions is equal to f. The secondary interaction appears in the form of a large repulsive force when the two ions are brought together so that their outermost electron shells are tangential to each other.

In an ionic bond, both the positive and negative ions resemble inert gases: because there is a system of complete electron shells, the balance of charges will not change without the introduction of a considerable alteration in the energy state of the system. Thus, when two ions are brought into proximity, their energy content increases rapidly; the repulsive force acting between the adjacent electron shells increases to a large value. Accordingly, the total potential energy ϕ of ion x in a crystal of a substance which is formed by an ionic bond is given by

$$\phi = \phi_M + \phi_R \tag{1.2}$$

where ϕ_M is the energy due to attractive interaction with all other ions in the crystal, and ϕ_R is the repulsive energy.

The attractive interaction may be expressed as

$$\phi_M = -\frac{z^2 e^2 A}{r} \tag{1.3}$$

where e is the electron charge, r is the distance between the centres of the closest pair of negative and positive ions, A is a constant (the Madelung number), and z is the number of charges.

The repulsive energy may be expressed as

$$\phi_R = \frac{B e^2}{r^n} \tag{1.4}$$

where B and n are constants (n is usually of the order of 9). If a crystal of one gram-molecular weight is considered, instead of a pair of ions, Equation 1.2 may be written as

5

$$U = -\frac{Nz^2e^2A}{r} + \frac{NBe^2}{r^n} \qquad (1.5)$$

where U is now the total lattice potential energy and N is Avogadro's number. In Equation 1.5, the only variable is r. Hence the minimum lattice potential energy may be expressed as

$$U_0 = -\frac{Nz^2e^2A}{r_0} + \frac{NBe^2}{r_0{}^n}$$

The variation of lattice potential energy with the distance between the centres of the closest pair of ions is shown in Fig. 1.1.

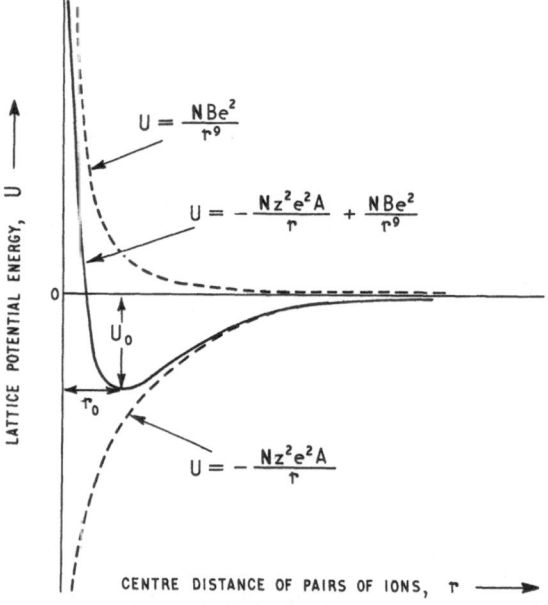

Fig. 1.1. Variation of lattice potential energy for ionic crystals

The ions within a theoretically perfect crystal will arrange themselves at a minimum energy level. It can be shown by simple differential calculus that the minimum energy state U_0 corresponds to a specific value r_0, which is thus the ionic distance at which the ions must settle in order to yield the minimum energy condition. As this r_0 dimension does not show directional characteristics, it can be proved that r_0 is repetitive along the coordinate directions within the boundary of the perfect crystal — hence the conclusion that the ions are arranged in an orderly and repetitive manner (the premise of the crystalline state).

Substances which form ionic crystals are hard and brittle, owing to the large energy variations relating to small changes in r.

1.2.2 MOLECULAR CRYSTAL BOND

In this group, the internally strongly bound neutral atomic clusters are held in position relative to each other by weaker van der Waals' forces. The forces that hold these types of solids together arise from a non-symmetrical charge distribution. The weak cohesive forces indicate low hardness, comparatively low melting points, and brittleness. Examples of substances with this molecular structure are iodine, arsenic trioxide, and urea.

1.2.3 COVALENT CRYSTAL BOND

The binding force of the crystal structure is based primarily on the equilibrium of the attractive and repulsive energies of the atoms composing the crystal. The main difference between ionic and covalent crystals is that the latter are bound together by electrons shared by two or more atoms. The type of structure is characteristic of elements which have four valency electrons: for example, carbon, silicon, germanium and grey tin.

Covalent crystal structures are also found in some intermetallic compounds, such as gallium arsenide, zinc selenide, and indium antimonide. The characteristic physical properties of the class are great hardness, high melting points, and brittleness.

In addition, covalent crystals follow the Hume–Rothery $8 - N$ rule, where N is the number of valency electrons and the factor $8 - N$ gives the number of nearest neighbours in the structure.

1.2.4 METALLIC CRYSTAL BOND

Metals are elements that have loosely bound valency electrons, which are free to move within the element. The elementary idea of a metal is that it consists of an array of positive ions immersed in

7

a cloud of negatively charged electrons. The valency electrons are assumed to move within the metal, in all directions, at high velocities. The binding force which holds metallic crystals together originates from the attraction of the ions for the surrounding electrons.

Because of the high mobility of the electrons forming the cloud, there is always a large number of electrons momentarily not participating in the formation of inter-ionic bonds; hence, these are free to carry charge from one place to another. The properties of high thermal and electric conductivity of metals are derived from the presence of free electrons. As the metallic crystal undergoes deformation, the distances between the ions are made either smaller or larger. If the distances are made smaller by compression, then the velocities of the free electrons increase, with a corresponding increase in their kinetic energy. This in turn produces a repulsive energy force, which becomes progressively larger. When the inter-ionic distance is increased by placing the crystal under tension, some of the free electrons aid the electron already providing the bond, thus maintaining the attractive forces between ions. This provides a simplified explanation of the elasticity and plasticity of metallic crystals.

1.3 CRYSTAL SYSTEMS AND THE BRAVAIS LATTICES

In Section 1.2.1, some reasons for the preference of ionic materials for solidifying in a crystalline pattern were considered. This crystalline preference may similarly be demonstrated for molecular, covalent and metallic crystals.

In many crystalline compounds, it may be easier to recognise the repetitive pattern if the single ions are replaced by a cluster of atoms as the unit pattern of recurrence. A suitable cluster is chosen, so that it reproduces the complete crystal by simple translatory movements along orthogonal, or near orthogonal, three-dimensional axes. The result of such a series of operations is a *space lattice*, in which every unit pattern has identical surroundings. The centre of gravity of the group of atoms selected as the unit pattern is often referred to as the *lattice point*.

Fig. 1.2(a) shows a group of atoms of two elements, P and Q. The molecule is formed in such a way that one P atom attracts two Q atoms; hence the chemical notation of this substance is PQ_2. If the group PQ_2 is translated a distance a along the x axis and a distance b along the y axis, parallel to the principal directions shown in Fig. 1.2(a), a two-dimensional lattice is produced [Fig. 1.2(b)]. If the operation is now extended a distance c in the direction of the

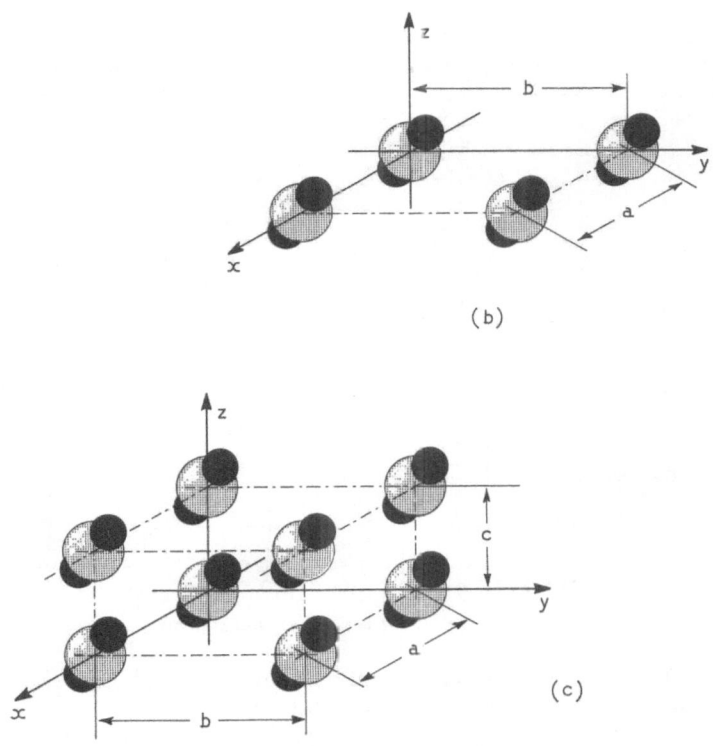

(b)

(c)

Fig. 1.2. Derivation of a space lattice

z axis [Fig. 1.2(c)], the group PQ_2 is located at the apices of a parallelepiped whose sides are of lengths a, b and c, and whose inter-axial angles are α, β and γ.

If similar parallelepipeds are stacked side by side in the x and y directions, and top to bottom in the direction of the z axis, an ideal crystal can be reproduced with lattice parameters a, b and c in the directions of the x, y and z axes respectively, and lattice angles α, β and γ between the lattice directions y and z, z and x, and x and y respectively.

9

It may be seen by inspection that each PQ_2 group of atoms has, in this arrangement, identical surroundings. The parallelepiped which is used for reproducing the space lattice is called the *unit cell*. It follows from the premise of identical surroundings of all

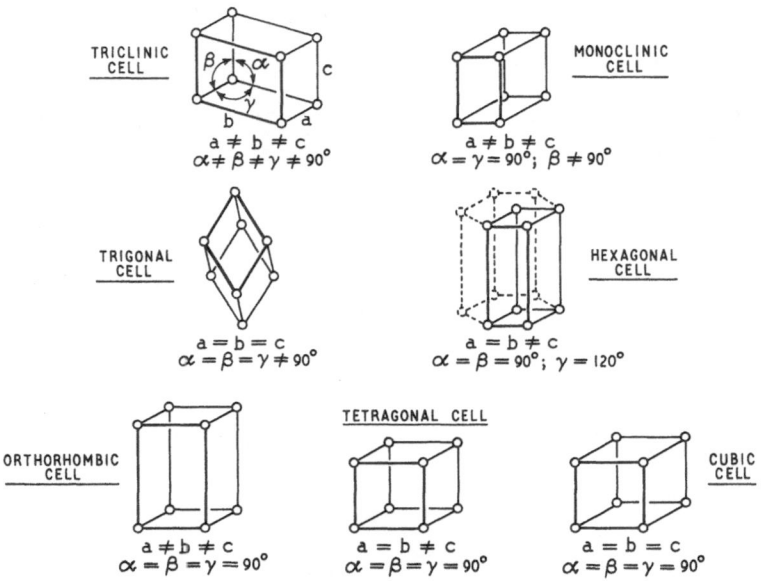

Fig. 1.3. The seven crystal systems

lattice points, that each unit cell in a crystalline substance is identical in size, shape and orientation.

These unit cells are, in fact, regarded as the building blocks of crystals – the relationship between unit cells and a crystal being comparable to that between the bricks and a completed building. As the relatively simply shape of the brick does not limit the complexity and expression of the building, in the same way, the simple shape of the unit cell does not limit nature in producing the most beautiful combinations of shapes of crystals.

The crystalline solids may be grouped into seven crystal systems based on the geometry of their unit cell. Consider a parallelepiped unit cell of dimensions a, b and c, and inter-axial angles α, β and γ. The most general – and therefore the least symmetrical – shape is derived when all the angular and length parameters are unequal,

10

and none of the angular parameters is a right angle. A unit cell of this description is called a triclinic cell. By successive geometrical simplifications to more symmetrical shapes, it may be shown that there are only seven different unit cell shapes possible in the whole field of crystals. It follows, therefore, that all crystalline matter may be classified into one of the seven systems of unit cell shape. The seven crystal systems are illustrated in Fig. 1.3, together with the generating parameters.

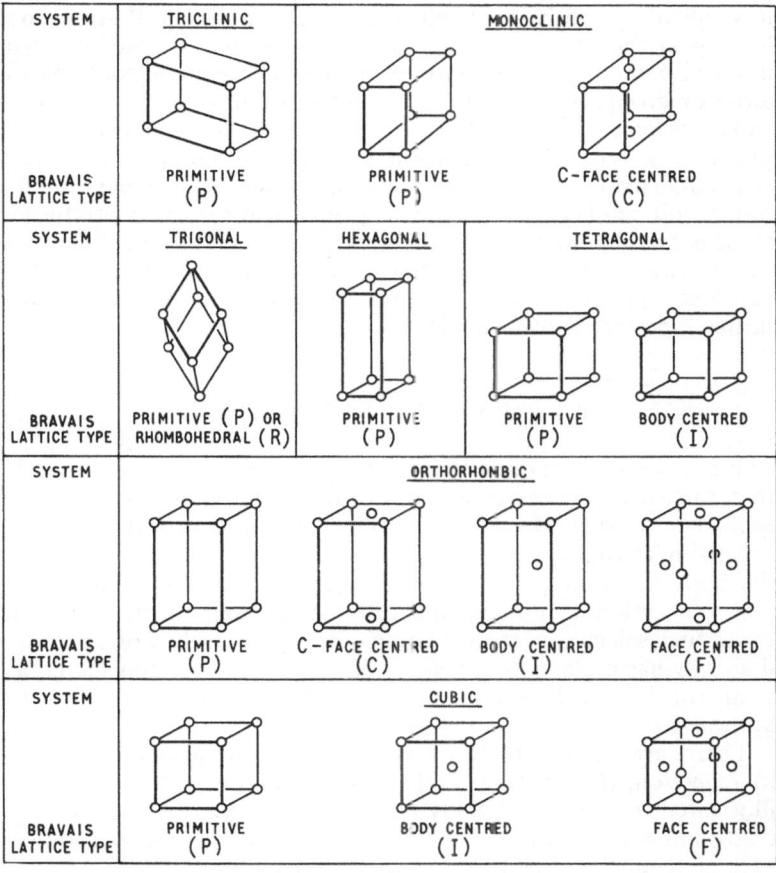

Fig. 1.4. The 14 Bravais lattices

If the geometrical reasoning is now extended to the examination of the atomic grouping within the unit cell, it may be shown that there are just 14 possible combinations to be found in space lattices. The classification was formulated by Bravais (1811–63), and these 14 space lattices are called the *Bravais lattices*. The Bravais lattices are deduced from the crystal systems in such a manner that a unit cell may contain an atom (or an ion, or a group) at the centre of its volume, or at the centre of all or some of its faces, providing the characteristic symmetry of the system is preserved. If the unit cell's atoms are arranged at the apices only, it is called a primitive unit cell of the relevant system. If atoms are located at the centre of unit cell faces, it becomes a face centred unit cell. Similarly, if the centre of the unit cell volume contains an atom or group, it is called a body centred unit cell. In the monoclinic system, for example, the centring of two opposite faces produces a type of arrangement known as a monoclinic *C*-face centred space lattice. If all the faces are centred, or if the volume of the unit cell is centred, no new geometrical system is produced, for it is possible to choose a differently orientated monoclinic unit cell in which only the two opposite faces are centred. In the ortho-rhombic system, all four variants are present. The 14 possible lattice types are shown in Fig. 1.4.

1.4 CRYSTAL SYMMETRY

The geometrical examination of the seven crystal systems reveals that each class possesses a certain minimum symmetry, which is exclusive to that class and therefore exclusive to the crystals belonging to that class. A body is said to have symmetry if parts of the body (for example, faces on a crystal) are arranged in such a way that certain symmetry operations performed on the body will bring that selected part into coincidence with itself. For instance, if a body has a plane of symmetry, the plane may be considered as a mirror in which a reflection of precisely half the body has appeared.

There are, in all, three morphological symmetry operations: (*i*) reflection, (*ii*) rotation, and (*iii*) rotation–inversion. These are illustrated in Fig. 1.5. A body has rotational symmetry if rotating it around an axis brings it into self-coincidence. These rotation axes of symmetry may be many-fold; in general, however, they are one, two, three, four and six-fold. (These are sometimes referred to as mono, diad, triad, tetrad and hexad axes.) For example, a four-fold axis of symmetry means that a body may be brought into congruent positions four times in one complete revolution of 360° around the axis of symmetry.

A body is said to have a rotation–inversion axis (which, like the rotation axes, may be many-fold) if equivalent points of the body

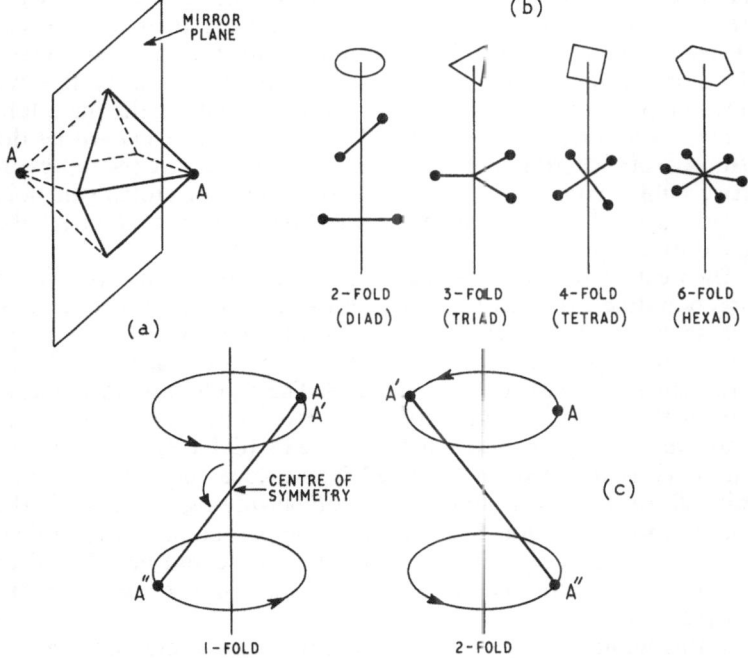

Fig. 1.5. Diagrammatic representation of the three morphological symmetry operations: (a) mirror or reflection symmetry; (b) rotation symmetry; (c) rotation–inversion symmetry

lying on a straight line may be brought into congruence by a rotation followed by an inversion around a centre lying on the axis. Crystals possessing such centres of symmetry are grouped as centro-symmetrical crystals. The detection of the presence or the lack of centro-symmetry in a substance of unknown crystal structure accelerates the process of structure determination very considerably.

It is now possible to re-examine the relationship between the crystal systems and Bravais lattices. The Bravais lattices are derived from the seven crystal systems by the addition of either

13

face centring or body centring atoms, in such a way that the characteristic symmetry of the system is retained. For example, the characteristic symmetry of the cubic system is the four triad axes of symmetry passing diagonally through the centro-symmetrical apices. The four triad symmetry axes are, in fact, the only symmetry element belonging exclusively to the cubic class of crystals. The four-fold symmetry axes present in the cubic class are also present in the tetragonal class and are therefore not exclusive to one class only. The two-fold symmetry axes and mirror planes are also shared with other less symmetrical classes. If, now, only one pair of mirror faces is considered as centred, the three-fold symmetry is lost, and the unit cell may be suitably chosen as the primitive of the tetragonal system. In order to preserve the representative cubic symmetry, extra atoms or groups of atoms can only be added to the primitive cell in the body centring or in all the face centring positions.

The next symmetry class of crystals is the tetragonal class, whose representative symmetry is one, and only one, four-fold axis of symmetry parallel to the c axis. Consider putting two atoms in face centring positions, one each in two opposite faces which are perpendicular to either the a or b axis. The result is the disappearance of the four-fold symmetry. If the two atoms are used to face centre the faces perpendicular to the c axis (C faces), the four-fold symmetry is retained and the cell is a C-face centred tetragonal unit cell. From one of the previous configurations, for example the B-face centred tetragonal unit cell (i.e., the pair of faces perpendicular to the b axis are centred), a new unit cell could be selected to become C-face centred and hence satisfy the symmetry requirement of the tetragonal class.

In this manner, all geometrical combinations could be tried and eventually it would be found that there are only 14 distinctly different lattices possible — those which are listed as the 14 Bravais lattices.

1.5 MILLER INDICES AND HEXAGONAL INDICES

In crystallographic work, it is often necessary to define planes and directions relative to a chosen coordinate system. For this purpose, the Miller indices are universally accepted.

A coordinate system, as shown in Fig. 1.6, is placed so that the

14

origin O coincides with the centre of symmetry of an orthorhombic unit cell. The interatomic distances along the x, y and z
axes are chosen as a, b and c respectively. The positive directions
are marked on the axes by arrows. Assuming that the coordinate
axes intersect plane m at points A, B and C, then triangle ABC is
contained in plane m. If now the positions of A, B and C are

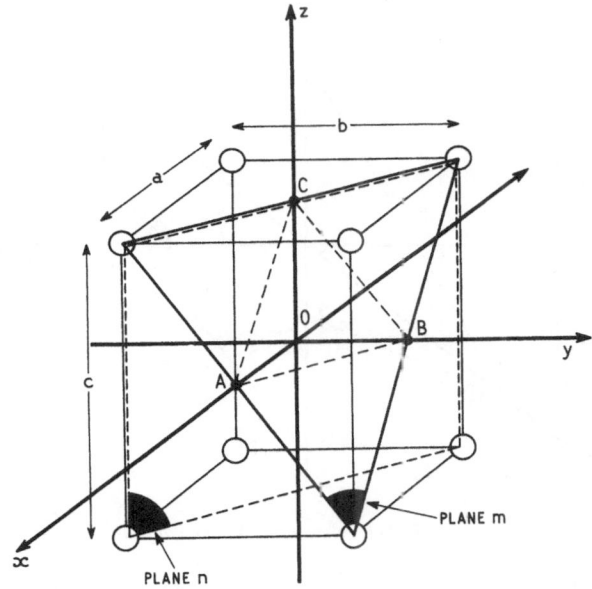

Fig. 1.6. The orthorhombic unit cell

defined, they will also define the orientation of plane m relative to
the assumed coordinate system. In the orthorhombic crystal
system, for example, the lattice parameters (interatomic distances)
are a, b and c, as shown in Fig. 1.6. The distances between the
origin O and the points of intersection may be expressed in terms
of the lattice parameters as follows:

$$OA = \tfrac{1}{2}a$$
$$OB = \tfrac{1}{2}b$$
$$OC = \tfrac{1}{2}c$$

By definition, the Miller indices of plane m are the reciprocal of the intercepts measured in terms of the lattice parameters and converted to the smallest possible integers. The reciprocals of the intercepts of Fig. 1.6, measured in terms of the lattice parameters, are 2, 2 and 2. To convert them into the smallest integers, the intercepts are divided by two; the Miller indices of plane m then

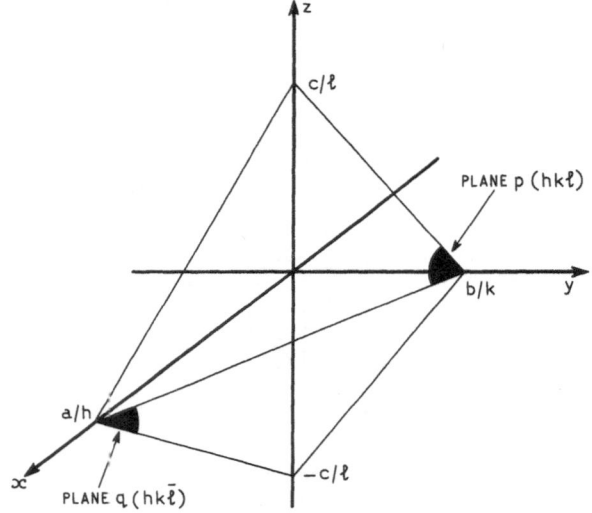

Fig. 1.7. The generalised Miller notation

become 1, 1 and 1, and plane m is called the one–one–one plane. To avoid ambiguity, the index values are printed within brackets thus (111), indicating a specific plane whose indices are one–one–one.

If a plane is orientated so that it becomes parallel to, say, the z axis, then the intercept OC becomes infinitely large, hence its reciprocal is zero. The Miller indices of plane n of Fig. 1.6 are therefore (110), and plane n is thus called the one–one–oh plane. Similarly, other planes parallel to the x axis and to the y axis will have indices (011) and (101) respectively.

To generalise the problem even further, consider a plane p which intersects the x, y and z axes at a/h, b/k and c/l respectively, as shown in Fig. 1.7. The Miller indices of plane p are then (hkl), and plane p is called the aitch–kay–ell plane. If the plane is orientated so that the intercept falls in the negative direction of one or more axes, the intercept (and hence its reciprocal) is negative. The Miller indices of plane q (Fig. 1.7) are then $(hk\bar{l})$ (aitch–kay–minus ell). The reader may verify that the chosen coordinate axes need not be perpendicular to one another for the above principle still to be valid.

Fig. 1.8 illustrates some crystals of the cubic systems showing the Miller indices of the representative faces. A closer examination of the Miller indices of the cube shows that each of its faces is parallel to two axes; therefore, the indices are the permutations of 1, 0 and 0. The (100) plane is equivalent to the ($\bar{1}$00) plane, as they are parallel, and the minus sign above means that the plane is located at the far side of the cube. For these reasons, the cube is said to consist of planes of a 'form' and is indicated by the use of braces: thus, the cube is of the form $\{100\}$. Similarly, as illustrated in

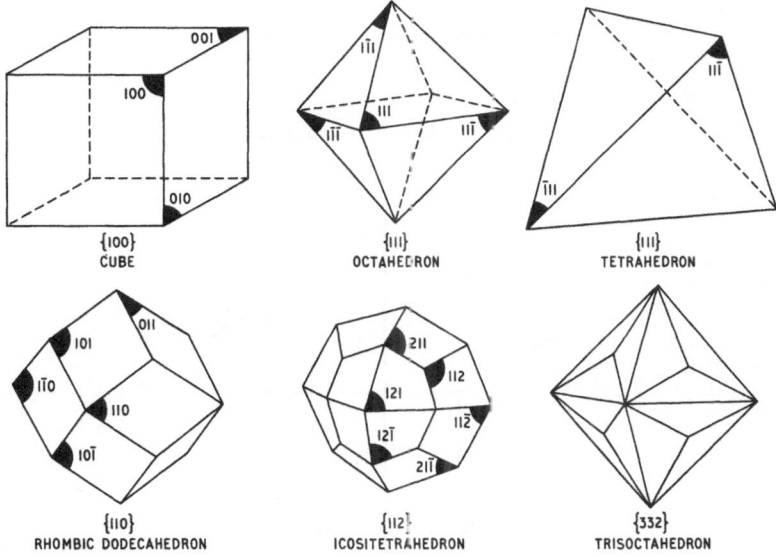

Fig. 1.8. Some crystal shapes of the cubic system, with indexed crystal faces

Fig. 1.8, the rhombic dodecahedron is of the form $\{110\}$, and the octahedron and the tetrahedron are of the form $\{111\}$.

The indices of directions are derived from simple vectorial considerations. Consider a point P in space, as illustrated in Fig. 1.9. Let the origin of the coordinate system be fixed at O. The direction

17

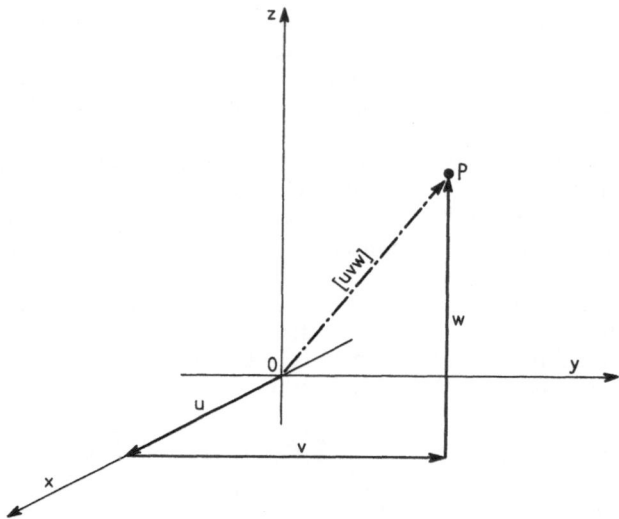

Fig. 1.9. The indices of direction

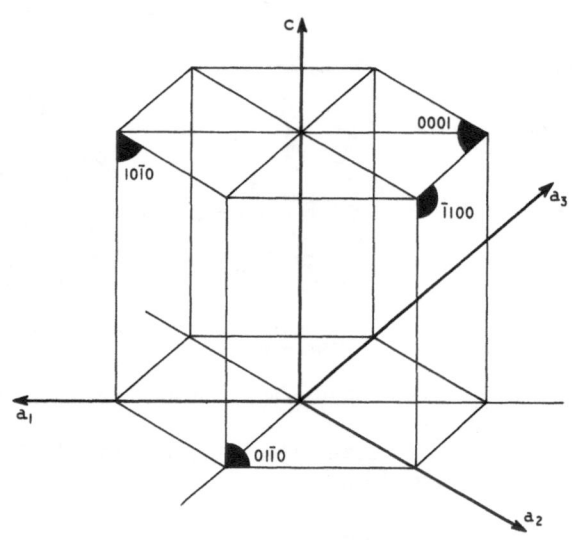

Fig. 1.10. The hexagonal indices

OP may be defined by a vector OP, or by the sum of the component vectors parallel to the principal (axial) directions. To reach point P from O, one must move a distance u in the x direction, a distance v in the y direction, and a distance w in the z direction. When u, v and w are converted into integers, they are called the indices of the direction OP and are denoted by square brackets $[uvw]$. The equivalent directions within the crystal are indicated by angle brackets $\langle uvw \rangle$.

The representative planes in the hexagonal system are often defined by four indices instead of the usual three. In this system there are four crystallographic axes: three co-planar axes a_1, a_2 and a_3 (Fig. 1.10), with 120° spacing between them, and one axis c at right angles to the plane. When the indices are computed in the usual way, they will appear in the form $(hkil)$. It can be proved that the indices h, k and i are always related by the equation:

$$i = -(h+k)$$

This system has the advantage that equivalent planes are represented by permutations of the h, k, and i indices only, the l index remaining the same. It may be easily verified, for example, that both the $(\bar{1}100)$ plane and the $(0\bar{1}10)$ plane are faces of a hexagonal prism and, therefore, equivalent; however, their equivalence is not self-evident from the usual Miller symbols. The hexagonal indices are often referred to as the Miller–Bravais indices, in memory of Auguste Bravais who extended the Miller system.

1.6 THE SIMPLE CRYSTAL STRUCTURE OF METALLIC ELEMENTS

About three-quarters of the elements have a number of common physical properties which are, to some degree, absent from the others. Such properties are high electrical and thermal conductivity, together with opacity in liquid and solid states. The elements possessing these properties are regarded as metals.

The first successful hypothesis about the metallic state was postulated by Drude and Lorentz at the beginning of this century. Under this hypothesis, the metallic elements are assumed to have loosely held valency electrons and the positive metallic ions to be immersed in a cloud of free electrons. The electromagnetic attraction between the ions and the surrounding free electrons provides the structural coherence, and the general mobility of the electron cloud provides the explanation for the ductility and conductivity of the metallic elements.

The presence of free electrons in metals has been demonstrated directly by R. C. Tolman. If the metal contains free electrons and is a conductor, then not all the free electrons are involved in maintaining the structural coherence; there will be some that can move under physical influence. As an electron has a mass of $9 \cdot 109 \times 10^{-28}$ g, if the electrons are accelerated and then stopped rapidly in accordance with the laws of classical mechanics, the free electrons, owing to their inertia, will continue to move within the metal.

Tolman found that after a rapidly vibrating metallic bar was stopped, the charged particles continued to move in the direction of the vibration. The presence of the momentary electrical potential difference, caused by the concentration of charges at one end of the bar, has been demonstrated by a gold leaf galvanometer connected to the ends of the bar. The negative nature of the moving charge has also been demonstrated, and the mass and charge of the moving particles measured within the limits of experimental error. Tolman showed the specific charge of the electrons to be $1 \cdot 77 \times 10^8$ C/g. This value has been found to be correct by other scientific workers using different experimental techniques.

The Drude–Lorentz theory of metals enables some conclusions to be drawn on the probable arrangement of ions making up the crystal lattice. The electromagnetic attraction between the ions and the surrounding electron cloud assumes neither a direction-dependence nor a numerical restriction on the metallic bond mechanism; hence the electromagnetic attraction between the ions, through the electron cloud, is regarded as spherically distributed and acting on as many neighbours as can be arranged around an ion. As the bonding forces are non-directional and unlimited in number, the crystal lattice of metals is henceforth dependent entirely on geometrical considerations.

Consider an ion, represented as a sphere in Fig. 1.11, surrounded by other similar ions. As the bonding forces are equidirectional, the surrounding ions tend to attain the greatest possible density around the centre ion A [Fig. 1.11(a)]. If the ions are of the same element, their representative spheres must be of uniform size, and hence the greatest number able to surround the centre ion (in two dimensions) is six. As the ions of the same metallic element are all precisely the same in size and in charge, any one ion may be chosen as the centring ion. If three more ions labelled B, are added to the group, C may also be chosen as a centring ion. If this two-dimensional lattice is extended to infinity, each ion equally becomes the centre of a group of six.

The close packing of ions may now be continued in the second layer, to build up the three-dimensional lattice of the metallic crystal, as shown in Fig. 1.11(b). This can be done by placing

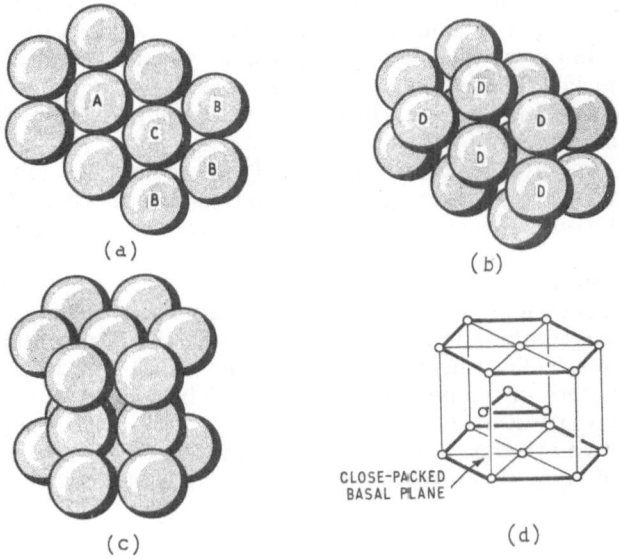

Fig. 1.11. The close-packed hexagonal structure: (a) and (b) stacking sequence; (c) packing diagram; (d) unit cell

spheres D into the depressions (or saddles) occurring between any three ions of the first close-packed plane. It is obvious from the geometrical construction that the ions of the second layer may only be placed into every alternate saddle of the first layer. This may be done in two ways, depending on the choice of the site for the first ion – although the choice of another centre shows that there is no fundamental difference between the two.

Now consider the second ionic layer on its own. It can be seen from Fig. 1.11(b) that a third ionic layer may be added to the second in two distinctly different ways: each ion can be placed either directly above a sphere of the first layer, or directly above an unfilled saddle of the first layer. The result of the former operation

21

is the *close-packed hexagonal* (CPH) structure shown in Fig. 1.11(c) and Fig. 1.11(d). This structure has a layer pattern of the type ABABA . . .

This hypothetical picture of a close-packed hexagonal crystal is reasonably near to the physical reality. If the base of the hexagonal unit cell is a and its height is c, and if the composing ions are assumed to be touching each other and to be perfectly spherical, it may be shown that the c/a ratio is 1·633. In fact, measured c/a ratios vary between 1·58 and 1·89. This variation can be explained by the assumption that, for the elements which crystallise in the close-packed hexagonal system, the inter-ionic bond is slightly directional, and the ions in contact are not perfect spheres but slight ellipsoids.

If the alternative arrangement is followed and the ions of the third layer are placed in the saddle points left unoccupied by the second layer, the structure will not show the same periodicity displayed by the three-layer close-packed hexagonal stacking. To restore periodicity, it is then necessary to add a fourth layer of close-packed ions, similar to the first layer. The resulting structure from this combination is a cubic lattice of the face centred type, commonly called the *face centred cubic* (FCC) structure, with a layer pattern of the type ABCABCA . . . The face centred unit cell is shown in Fig. 1.12.

The two systems, both close-packed, are equally good from the point of view of stacking efficiency. The main difference between

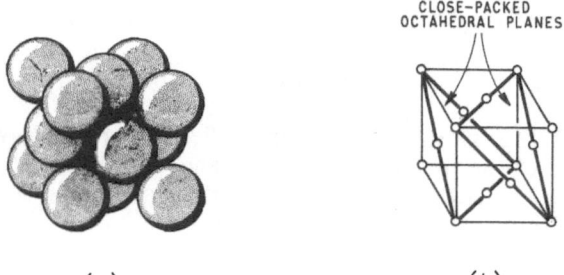

CLOSE–PACKED
OCTAHEDRAL PLANES

(a)   (b)

Fig. 1.12. A face centred cubic cell: (a) packing diagram; (b) unit cell

the systems is that the hexagonal system has only one plane (the basal plane) in which the individual ions are close-packed, whereas the face centred cubic structure possesses four such planes (octahedral planes).

22

The third most commonly occurring crystal lattice is called the *body centred cubic* (BCC) structure, shown in Fig. 1.13. This is again modelled on economical packing of round spheres which represent the metallic ions. Assuming that the metallic ions form equidirectional bonds and also that the unit cell must stack

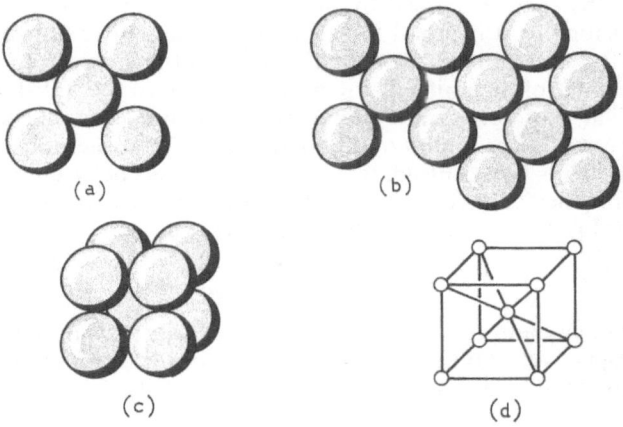

Fig. 1.13. *A body centred cubic cell: (a) and (b) stacking sequence; (c) close-packed unit cell; (d) open type of unit cell*

to fill space, it follows that the ions will need to arrange themselves in a space symmetrical pattern. Such a space symmetrical arrangement is the body centred cubic structure.

The stacking efficiency of a unit cell may be described by its coordination number. The coordination number of a unit cell is equal to the number of nearest neighbours that any one ion possesses in the lattice structure.

A study of the face centred cubic and the close-packed hexagonal unit cells shows that any one ion is in intimate contact with 12 others. Thus the coordination number of both systems is 12. It can also be deduced by simple mathematics that both systems are equally efficient in filling space. If the diameter of each ion is d, than the volume of the unit cell per ion is in both cases $0.7075\ d^3$; the percentage of the unit cell space which is occupied by the ions is 74.1%.

In the body centred cubic lattice, the body centring ion has no special significance over the corner ones, since all the ions of the lattice are equivalent. The simple inspection of Fig. 1.13 shows that the body centring ion is in intimate contact with eight others, hence its coordination number is eight. The volume of the unit cell per ion is $0.770\ d^3$, and the percentage of the unit cell space occupied is 68.1%. The difference in stacking efficiency between the close-packed and the body centred cubic structures is relatively small. The body centred cubic structure occupies approximately 92% of the space occupied by the close-packed structure.

Occasionally, it is useful to calculate the number of ions belonging to the unit cell. The ions which are contained entirely within the precincts of the unit cell will wholly belong to the cell. The ions located at the corners are shared by eight unit cells, hence their contribution is one-eighth of the total number of corner ions. When the unit cell is face centred, these ions are shared between two unit cells, thus they contribute half of their total number to the unit cell. The total number of ions may be expressed as

$$N = N_i + \tfrac{1}{2}N_f + \tfrac{1}{8}N_c$$

where N_i, N_f and N_c are the number of interior, facial and corner ions respectively. Of the 83 representative and transitional metallic elements, 58 display (in one or more stages of their existence) one

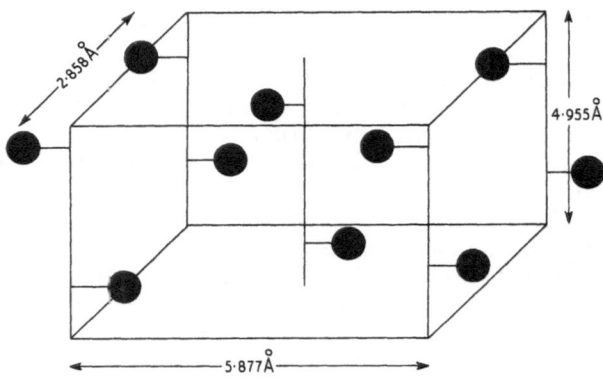

Fig. 1.14. The structure of uranium

of the three structures discussed above; the other 15 crystallise in a more complex lattice structure, such as the rhombohedral, tetragonal or orthorhombic structures.

The geometrical principle of crystallisation of metallic elements may be postulated thus: the metallic ions of a crystal are located in

24

space on the points of a Bravais lattice or in some geometrical relationship to those points. For example, in the structure of uranium (Fig. 1.14), the outline of the unit cell can be chosen so that it is recognisable as a C-face centred orthorhombic cell. α-Uranium is an example of the association of more than one ion with each Bravais lattice point. In this instance, the unit cell contains four ions; however, it should be apparent that the number of ions combined within the unit cell is not limited. In complex organic structures, the unit cells in some instances contain more than 100 atoms.

1.7 THE SIMPLE STRUCTURE OF ALLOYS

Most engineering applications of metals require some very complex and exacting physicochemical properties which are either not possessed by the metallic elements or only to a limited extent. Even the earliest metallographers realised that the properties of the Cu–Zn alloys (bronzes) were superior in many ways to those of the component elements. Alloys are combinations of metallic, non-metallic and metalloid substances in solution which nevertheless retain their metallic properties. Alloys composed of two, three or four elements, are named binary, ternary and quaternary alloys respectively. Considering that there are about 70 metallic elements, their binary combinations result in 4,830 systems of which about 1,000 combinations have been thoroughly examined. The ternary combinations number 328,440 systems. It is not necessary to emphasise that a great deal of further research is required for the full understanding of the metallic systems.

Commercial alloys normally contain more than two components, but it is often possible and sufficient to represent them in a binary form of the two principal elements; the other non-representative elements are treated as agents to modify or enhance some properties already present in the binary system. In this section, consideration is given to binary systems only.

Alloys are generally formed in the molten state and are then allowed to cool, passing through the solidification phase, to room temperature. Some metals nevertheless form no solution in either the liquid or the solid state. For example, iron forms neither a liquid nor a solid solution with silver, cadmium or bismuth, among many others. In such instances, the two metals exist at all temperatures as separate phases, side by side.

Some pairs of metals can form a complete solution at all temperatures. For example, unlimited solubility is found in Ag–Au, Au–Pd, Cu–Ni and Bi–Sb systems. The crystal lattice of one of these systems consists of atoms of the two metals in equivalent positions. In other words, atoms of one element will replace atoms in the crystal lattice of the other element. A solution of this description is termed a *substitutional solid solution*. In the majority of cases, however, there is limited solubility between elements; also, the solubility depends, to a large extent, on the temperature. For example, silver can keep 1·3 atomic % of copper in solution at 250°C, decreasing to about 0·2 atomic % at room temperature. In reverse, copper may contain 2·7 atomic % of silver at 700°C, decreasing to 0·1 atomic %, at room temperature.

The formation of a substitutional solid solution is governed by the Hume–Rothery law, which states that, if a continuous solid solution between two elements is to be obtained:

1. The two elements must crystallise in the same lattice structure.
2. Their atomic sizes must differ by no more than 14%.
3. Their electrochemical properties must be similar.

When these three conditions are fulfilled, the solvent atoms replace the solute atoms in a random distribution. For instance, silver and gold both crystallise in a face centred cubic lattice structure; their atomic sizes differs by less than 0·2%; and both metals belong to the same column of the periodic table.

The Au–Ag system, in fact, forms a substitutional solid solution in the liquid and in the solid state at all temperatures and compositions. It crystallises in the face centred cubic structure, and the lattice parameter changes approximately linearly with composition. This change in lattice parameter is caused by the distortion of cells in which substitution takes place. Although X-ray powder photographs show sharp diffraction lines, they do not prove that all cells are similar in size. The interpretation of powder photographs yields information on the average cell dimension, but it does not allow for large local variations.

Solubility also occurs between metals of different chemical properties. Such solubility is governed by two factors. The first factor is the electro-negative valency effect, by which the relative potential difference between the atoms of the solvent and solute elements restricts the range in which a solid solution forms and which generally encourages the formation of intermetallic compounds. The second factor is the relative valency effect, by which the metal of lower valency dissolves a metal of higher valency more readily than vice versa. For example, magnesium of valency 2 dissolves less than 0·2 atomic % of gold of valency 1, but gold dissolves 30 atomic % of magnesium.

The other type of substitutional solid solution comes about by the orderly arrangement of the solute atoms in the solvent's lattice structure. Solutions of this type are termed *ordered substitutional solid solutions*, and the solution crystal is said to have a super lattice. A typical example of super lattice formation is the $AuCu_3$ alloys. At elevated temperatures, the copper and gold atoms are distributed statistically in a face centred cubic lattice; at room temperature, the atoms rearrange themselves so that the gold atoms occupy the corner positions of the cube while the copper atoms locate themselves at the face centring positions.

When the atomic sizes of the two elements in solution differ greatly, a new type of solid solution occurs. Since the solute atoms are not large enough to replace the solvent atoms on the lattice, they site themselves in the interstices between the spherical atoms. Hence, the solution is called an *interstitial solid solution*. There are only five elements (hydrogen, boron, carbon, nitrogen and oxygen) that are known to occupy interstitial lattice positions. It is estimated that, in the interstitial solution of carbon in α-iron (ferrite), the number of carbon atoms does not exceed one in 500 unit cells since the solubility of carbon in ferrite is limited to about 0·1 atomic %. It is possible to distinguish this type of solution from the combination of average lattice parameters and densities.

For example, suppose that a atomic % of a metalloid of atomic weight Y is dissolved in a body centred cubic metal of atomic weight X. The mass of the unit cell may be calculated for both the substitutional (M_{subst}) and the interstitial (M_{int}) solutions as

$$M_{subst} = \frac{2X}{N} + \frac{200aN}{Y-X}$$

$$M_{int} = \frac{2X}{N} + \frac{2aN}{100Y}$$

where N is Avogadro's number.

The volume of the unit cell may readily be calculated from the lattice parameters. If the calculated masses are divided by the volume, a theoretical density value can be determined. The comparison of these calculated values with results obtained experimentally should show the type of solid solution present.

27

2
Physics of X-rays

2.1 NATURE AND GENERATION OF X-RAYS

X-rays are produced when rapidly moving electrons of sufficient energy strike a target and are rapidly decelerated. The electrons are normally produced by thermal emission from a metallic wire, usually of tungsten; a large kinetic energy is provided by the application of a high potential difference to accelerate the electrons between the heated wire and a metallic target. The kinetic energy E of the electrons may be expressed (in ergs) by the equation

$$E = \tfrac{1}{2}mv^2 \qquad (2.1)$$

where m is the mass of an electron ($9\cdot109 \times 10^{-28}$ g) and v is its velocity, in cm/s, before impact. An alternative expression for the kinetic energy of an electron is

$$E = eV_{\text{esu}} \qquad (2.2)$$

where e is the electron charge ($4\cdot803 \times 10^{-10}$ e.s.u.), and V_{esu} is the potential difference, in e.s.u., between the source and the target.

Fig. 2.1 shows that, for each potential difference, there is a limiting wavelength λ_0 below which no X-rays are produced. The intensity I of the longer wavelengths increases with the increase in applied potential difference. According to the Planck–Einstein quantum equation, the kinetic energy of an electron is given by the equation

$$eV_{\text{esu}} = h\nu_0 = \frac{hc}{\lambda_0} \qquad (2.3)$$

where h is Planck's constant (6.625×10^{-27} erg s), v_0 is the threshold frequency corresponding to the limiting wavelength λ_0, and c is the velocity of light in vacuo (2.9979×10^{10} cm/s). Rearrangement of Equation 2.3 gives

$$\lambda_0 = \frac{hc}{eV_{esu}} \tag{2.4}$$

If the known values of constants are substituted into Equation 2.4, and if 1 volt is taken as $\frac{1}{300}$ e.s.u. of potential difference, the limiting wavelength for a given applied voltage V becomes

$$\lambda_0 = \frac{12,403 \cdot 246}{V} \times 10^{-8} \text{ cm} \tag{2.5}$$

or, approximately,

$$\lambda_0 = \frac{12,345}{V} \text{ Å} \tag{2.6}$$

This last expression is known as the Duane–Hunt law, the number 12,345 providing a convenient mnemonic.

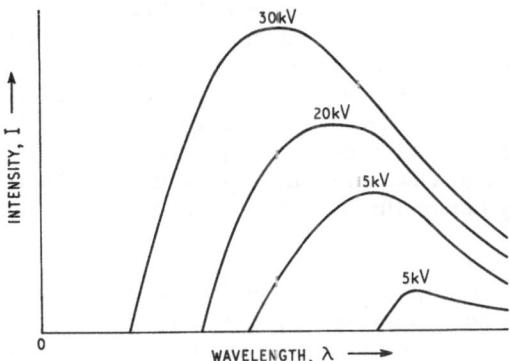

Fig. 2.1. The continuous X-ray spectrum for various potential differences

The kinetic energy E of an accelerated electron is dissipated when it collides with the electrons in the target, causing radiation of wavelength λ to be emitted. Not every electron is decelerated during a collision; also, an electron may emerge from one collision still retaining sufficient energy to go on to another collision—it will then lose a further quantity of energy by emitting a radiation of

29

lesser frequency (and consequently of longer wavelength). The radiation obtained in this way is continuous from the limiting wavelength to very long wavelengths. Since this type of radiation is made up of a variety of different wavelengths, similar to white light, it is often referred to as *white radiation*.

While the limiting wavelength of the continuous radiation is independent of the target material, the total intensity I_{total} of the radiation (the area under the intensity curve for a selected voltage) is a function of the atomic number of the target material and the operating current. This integrated intensity for an applied voltage V is given by the empirical expression

$$I_{total} = AiZV^2 \qquad (2.7)$$

where A is a constant of proportionality, i is the operating current in amperes, and Z is the atomic number of the target material. The intensity curve rises steeply to a maximum at about $1.5\lambda_0$ and then falls away asymptotically to zero intensity.

If the accelerating potential is increased above a certain threshold value, a characteristic line spectrum will appear, superimposed on the continuous effect. These lines were first discovered by C. G. Barkla, and are known as *characteristic radiation*.

The nature of characteristic radiation is different from that of white radiation, in that the accelerated electrons possess enough (or indeed an excess) of energy to knock electrons out of the electron shells of atoms of the target material. The minimum energy required to accomplish this process is the ionisation energy of the particular electron shell in question. The electron shells are traditionally named the K, L, M, N, ... shells, the innermost being the K shell.

When a K electron is knocked out, a vacancy is created in the K shell and the atom is left in an excited state. The electrons located in the outer shells possess ionisation energies in excess of that of K shell electrons. As the more energetic electrons of the outer orbits occupy the vacant site in the K shell, they emit a distinct quantum of energy in order to redress the balance of charges; this quantum of energy is equal to the difference between the ionisation energy of the outer shell and the ionisation energy of the K shell. A K shell vacancy may be filled by an electron from any one of the outer shells; on statistical evidence, however, only the three next shells provide electrons with useful short wavelengths to fill the vacancies—thus giving rise to three characteristic K lines: the $K\alpha$ line, the $K\beta$ line and the $K\gamma$ line.

A further distinction can be made between the spectrum lines, in that the probability of an *L* shell electron filling the *K* vacancy is greater than the probability of an *M* shell electron moving to the *K* orbit. The result is that the intensity of the *Kα* line is expected to be higher than that of the *Kβ* line; and the *Kβ* line is expected to be more intense than the *Kγ* line.

As the energy levels between the shells differ by discrete quantities, the intensity of the characteristic radiation is considerable

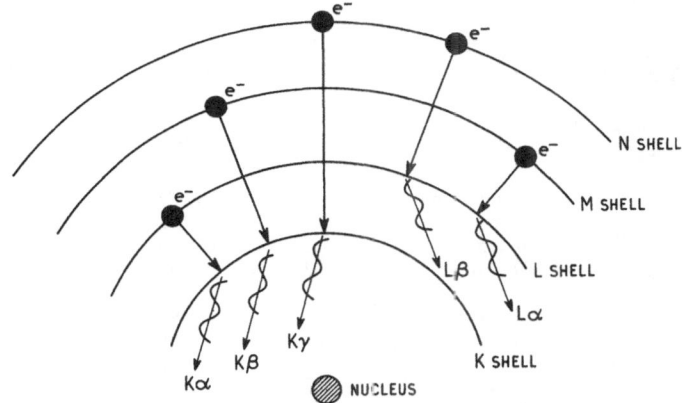

Fig. 2.2. *Origin of the characteristic line spectrum of X-rays*

and its wavelength is of a single value (at least as a first approximation) so that it appears as a line spectrum. Fig. 2.2 shows some selected electron orbits, and the associated transitions causing the appearance of characteristic radiation.

Further refinements in experimental techniques have shown the *Kα* emission spectra to be of two lines α_1 and α_2 (whose intensity ratio I_{α_1} to I_{α_2} is equal to 2 to 1). The existence of these two lines may be indicative of the small energy differences between the sub-orbits of electrons of the same principal quantum number.

The excitation voltage for any characteristic wavelength may be expressed by an equation similar to Equation 2.6. For instance, the excitation voltage V_K for *K* radiation is given by the equation

$$V_K = \frac{12,345}{\lambda_K} \tag{2.8}$$

where λ_K is the critical *K* absorption wavelength in Ångstroms.

31

Table 2.1 shows the critical K absorption wavelengths of some selected target materials.

The intensity I_c of the characteristic spectrum is given by the equation

$$I_c = Bi(V - V_K)^n \qquad (2.9)$$

where B and n are experimental constants, i is the operating current, and V and V_K are the operating and the K excitation voltages respectively.

Table 2.1.

The Critical K Absorption Wavelengths (λ_K) of some Selected Elements

Element	Atomic number	λ_K(Å)
Magnesium	12	9·5117
Aluminium	13	7·9511
Silicon	14	6·7446
Titanium	22	2·4973
Vanadium	23	2·2690
Chromium	24	2·0701
Manganese	25	1·8964
Iron	26	1·7433
Cobalt	27	1·6081
Nickel	28	1·4880
Copper	29	1·3804
Zinc	30	1·2833
Zirconium	40	0·6888
Molybdenum	42	0·6198
Silver	47	0·4858
Tin	50	0·4247
Tungsten	74	0·1784
Iridium	77	0·1629
Platinum	78	0·1582
Gold	79	0·1534

According to Guinier, the ratio of intensity I_K of K radiation to that of the total intensity I_{total} may be expressed by the equation

$$\frac{I_K}{I_{\text{total}}} = C(Z V_K^{0·5})^{-1} + \frac{[(V/V_K) - 1]^{1·5}}{(V/V_K)^2} \qquad (2.10)$$

where C is a proportionality constant, Z is the atomic number of the target material, and V and V_K are the operating and the K excitation voltages respectively.

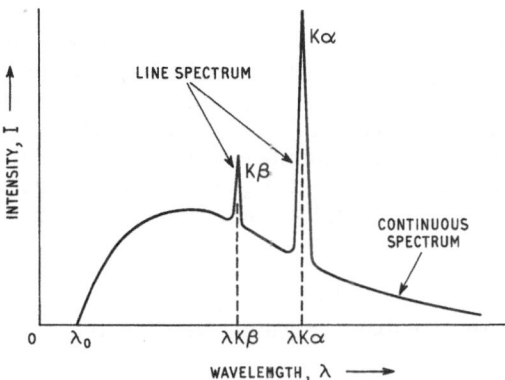

Fig. 2.3. Characteristic and continuous X-ray spectra

Fig. 2.3 shows diagrammatically the distribution of X-ray wavelengths in terms of intensities. The characteristic radiations appear as peak intensities over the continuous spectrum.

2.2 MOSELEY'S LAW

The brilliant work of M. G. T. Moseley greatly increased the scope of knowledge of atomic structure. He was the first to recognise the simplicity of the relationship between the atomic number and the multitude of physical properties of atoms

In his experiments, X-rays from various targets impinged on a crystal of potassium ferrocyanide. It was shown that the wavelength of a given characteristic spectral line increased continuously as the atomic number increased. In generalising this principle, Moseley derived his famous relationship: that the square root of the frequency v is proportional to the difference between the atomic number Z and the screening constant σ. The relationship can be written as

$$\sqrt{v} = k(Z - \sigma) \tag{2.11}$$

where k is a constant. This equation is known as Moseley's law. This law provided the key to the relatively recent discoveries of previously unknown elements, such as hafnium (Hf, $Z = 72$), technetium (Tc, $Z = 43$), and radioactive transformation products − neptunium (Np, $Z = 93$) and curium (Cm, $Z = 96$).

33

Moseley's work also presented the possibility of qualitative and quantitative analysis of unknown substances from their characteristic X-ray emission spectra. The frequency $v_{K\alpha}$ of the $K\alpha$ lines may be obtained from Equation 2.11 by substitution of values for k and σ. The evaluation of Moseley's equation is somewhat simplified by the introduction of wave numbers $\bar{v}$, which denote the number of vibrations per centimetre. If Equation 2.11 is squared and the values of the experimental constants k and σ are substituted, it becomes

$$v_{K\alpha} = \tfrac{3}{4}R(Z-1)^2 \tag{2.12}$$

where **R** is the Rydberg constant, which is equal to 109,737·323 cm^{-1}.

2.3 ABSORPTION OF X-RAYS

A low voltage applied between the electron source and target invariably results in X-rays of low intensity and long wavelength. These rays are not very penetrating and are said to be 'soft' rays. A high applied voltage normally improves the penetrating capacity of X-rays. The increased penetrating capacity is indicated by the name 'hard' rays. The absorption characteristics of matter are a function of the intensity and the wavelength, and they also depend

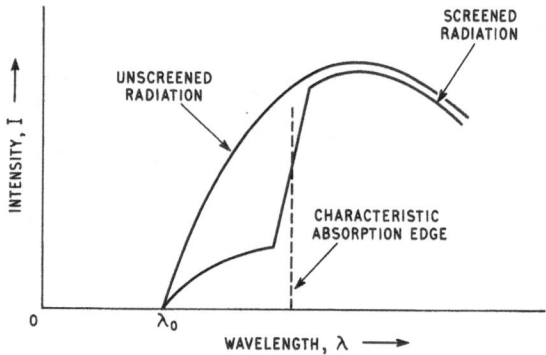

Fig. 2.4. Effect of screening on an X-ray spectrum

on the nature of the absorbing matter itself. Absorption occurs when the energy of the impacting electron is imparted to the

34

screening material, which then emits a photo-electron and increases its heat content.

As with characteristic radiation, there also exists a characteristic absorption spectrum, displaying discontinuities (absorption edges), for each of the elements. Discontinuity occurs because the energy required to remove a K shell electron from a given atom is discrete and governed by the quantum relationship $E = hv$. Each quantum of energy corresponds to a definite frequency (and hence to a definite wavelength), resulting in the K absorption edge. X-rays with wavelengths shorter than that of the absorption edge will appear as absorbent soft rays; while those with wavelengths above the absorption edge will pass through the absorbing screen with little loss in intensity. In Fig. 2.4, the variation in intensity is plotted against the wavelength passing through an absorbing screen.

The intensity of X-rays passing through homogeneous matter will diminish in proportion to the distance penetrated. Consider an X-ray beam of initial intensity I_0 which, on passing through a substance of thickness t, diminishes its intensity to I. Hence, a small change dt in the distance travelled results in a small change dI in intensity. Since the actual change in intensity for a constant thickness t is also dependent on both the wavelength of the travelling ray and the atomic number of the matter it penetrates, the differential equation representing the changing quantities must carry a factor which incorporates both variables. This constant of proportionality is called the linear absorption coefficient and is denoted by μ. The differential equation then becomes:

$$\frac{dI}{I} = -\mu dt \qquad (2.13)$$

The minus sign represents the decrease in intensity. Integration between the limits I_0 and I gives:

$$I = I_0 \exp(-\mu t) \qquad (2.14)$$

If the ratio μ/ρ (the mass absorption coefficient) is introduced, the fluctuations in absorption power due to density variations are eliminated. Equation 2.14 then becomes:

$$I = I_0 \exp[(-\mu/\rho)\rho t] \qquad (2.15)$$

Table 2.2 lists values of the mass absorption coefficient for some selected elements and for different radiations.

35

Table 2.2.

Mass Absorption Coefficients (μ/ρ) of Selected Elements for Various Radiations

Absorber	Atomic number	Radiation					
		Ag $K\alpha$	Mo $K\alpha$	Cu $K\alpha$	Co $K\alpha$	Fe $K\alpha$	Cr $K\alpha$
Nitrogen	7	0·60	1·10	8·51	13·6	17·3	27·7
Oxygen	8	0·80	1·50	12·7	20·2	25·2	40·1
Magnesium	12	2·27	4·38	40·6	60·0	75·7	120
Chromium	24	15·7	30·4	259	392	490	89·9
Iron	26	19·9	38·3	324	59·5	72·8	115
Copper	29	26·4	49·7	52·7	79·8	98·8	154
Molybdenum	42	70·7	20·2	164	242	299	439
Tin	50	17·4	33·3	265	382	457	681
Tungsten	74	54·6	105	171	258	320	456
Platinum	78	64·2	128	205	304	376	518
Gold	79	66·7	132	214	317	390	537

If the screen is composed of more than one element, the total mass absorption coefficient is the sum of the individual mass absorption coefficients $[(\mu/\rho)_1, (\mu/\rho)_2, ..., (\mu/\rho)_n]$ in proportion to their weights ($w_1, w_2, ..., w_n$ respectively) in the screening material. Thus:

$$(\mu/\rho)_{\text{total}} = w_1(\mu/\rho)_1 + w_2(\mu/\rho)_2 + ... + w_n(\mu/\rho)_n \qquad (2.16)$$

In Fig. 2.5, the variation of mass absorption coefficient with wavelength λ is shown. If the logarithmic values of μ/ρ and λ were plotted, the curved portions of the graph would become straight;

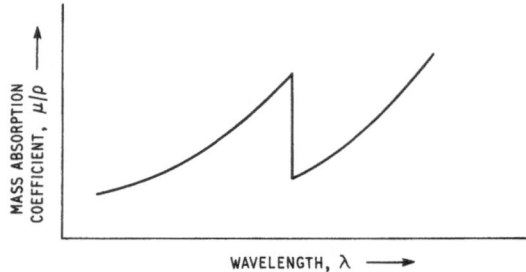

Fig. 2.5. Variation of the mass absorption coefficient with wavelength

the equations describing each straight line could then be written in the form

$$\log (\mu/\rho) = n \log \lambda + c' \tag{2.17}$$

where c' is a constant for each straight part of the diagram, and the value of n is found experimentally to be about 3. Therefore

$$\log (\mu/\rho) = \log \lambda^n + \log c$$

where $\log c = c'$. Therefore

$$\log (\mu/\rho) = \log c\lambda^n$$

or

$$\mu/\rho = c\lambda^n \tag{2.18}$$

2.4 FILTERING AND MONOCHROMATORS

In general metallurgical X-ray investigations, the heterogeneity of the radiation introduces an undesirable difficulty into the interpretation of diffraction data. This can be overcome considerably by the introduction of a suitable filter between the heterogeneous source of radiation and the crystal; this will eliminate, or at least reduce, the intensity of the unwanted radiation. A small amount of white radiation or the presence of the L or M series of characteristic radiation will not normally introduce uncertainties into the interpretation: the mixed wavelengths of white radiation will only affect the background darkening of the photographic plate, while the longer wavelengths of the higher order characteristic radiation will normally possess insufficient energy to produce a diffraction pattern.

The successful filtering out of $K\beta$ radiation depends on the absorption edge of the filtering material. In general, it should lie between the $K\alpha$ and $K\beta$ wavelengths of the target material. For the most extensively used copper radiation, the filter is made of nickel. The introduction of a nickel filtering foil reduces the intensity of the copper $K\beta$ radiation to about one six-hundredth of the $K\alpha$ intensity. Table 2.3 contains some selected data for the most useful filter materials suitable for a variety of radiations.

Substances which are impinged by X-rays will emit secondary X-rays restricted to a very narrow range of wavelengths. This property can be used to separate wavelengths if a single crystal is suitably positioned in a mechanical device. This separation of X-ray wavelengths by a single crystal is comparable to the dispersion of light by a refracting prism. The separating devices are called monochromators, and are manufactured either as plane or curved-crystal arrangements. Fig. 2.6 shows a typical plane-crystal monochromator.

Monochromation of similar quality may be obtained by means of a double filter technique, first developed by P. A. Ross. Two filters of slightly different thicknesses are chosen so that each of them absorbs a given wavelength equally. This is achieved when the mass absorption coefficient is the same for both filters. To ensure a

Table 2.3.

Selected Data for the Most Useful Target and Filter Materials

Target material	Atomic number	Emission wavelengths (Å)	Filter material	Atomic number	Absorption edge (Å)	Thickness (mm)
Chromium	24	$K\alpha_1 = 2\cdot28962$ $K\alpha_2 = 2\cdot29351$ $K\beta = 2\cdot08480$	vanadium	23	2·2690	0·016
Iron	26	$K\alpha_1 = 1\cdot93597$ $K\alpha_2 = 1\cdot93991$ $K\beta = 1\cdot75653$	manganese	25	1·8963	0·016
Cobalt	27	$K\alpha_1 = 1\cdot78892$ $K\alpha_2 = 1\cdot79278$ $K\beta = 1\cdot62075$	iron	26	1·7433	0·018
Nickel	28	$K\alpha_1 = 1\cdot65784$ $K\alpha_2 = 1\cdot66169$ $K\beta = 1\cdot50010$	cobalt	27	1·6081	0·018
Copper	29	$K\alpha_1 = 1\cdot54050$ $K\alpha_2 = 1\cdot54433$ $K\beta = 1\cdot39217$	nickel	28	1·4880	0·021
Molybdenum	42	$K\alpha_1 = 0\cdot70926$ $K\alpha_2 = 0\cdot71354$ $K\beta = 0\cdot63225$	zirconium	40	0·6888	0·108
Silver	47	$K\alpha_1 = 0\cdot55936$	rhodium	45	0·5337	0·079
		$K\alpha_2 = 0\cdot56377$ $K\beta = 0\cdot49701$	palladium	46	0·5091	

narrow band of wavelengths, the filtering elements should be adjacent in the periodic table. For copper $K\alpha$ radiation of wavelength 1·5418 Å, filters of nickel and cobalt should be used. The absorption edges are 1·4880 Å for nickel and 1·6081 Å for cobalt. Consequently the spread of copper $K\alpha$ radiation is closely channelled between the two absorption edge values. Since the gap

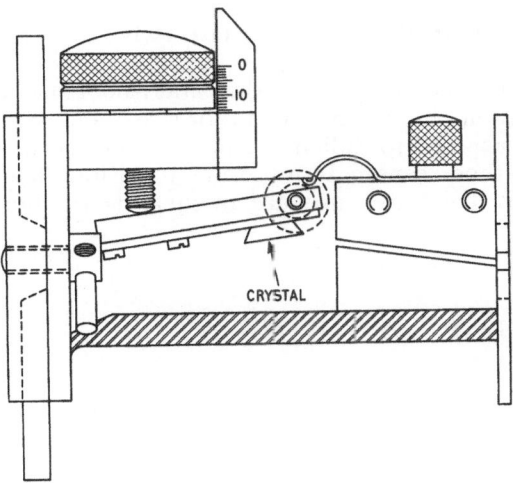

Fig. 2.6. Diagram of a plane-crystal monochromator. (Courtesy Unicam Instruments Ltd)

between the absorption edges is slightly wider than the spread of the copper $K\alpha$ doublet, a small amount of white radiation will still pass through the filters, increasing the background fog of the photographic plate.

The existence of absorption edge effects means that the wavelength of the radiation used for photographing a particular substance must be carefully selected. If the wavelength of the incident radiation is below that of the absorption edge, the diffraction effect will be weak, and heavily scattered (fluorescent) radiation will result. For example, the K absorption edge of chromium is 2·0701 Å, the wavelength of copper $K\alpha$ radiation is 1·5418 Å; hence copper radiation is unsuitable for specimens containing chromium. The intensity of the fluorescent radiation decreases as the gap between the wavelengths of the absorption edge and the incident radiation increases. The effect of scattered radiation may be greatly reduced by the introduction of suitable screens between the photographic plate and the specimen. For instance, if a nickel screen is placed between specimens of iron or chromium and the recording film, copper $K\alpha$ radiation can be successfully employed. Screening becomes extremely difficult when the wavelengths of the incident and excited radiations are close.

2.5 PHOTOGRAPHIC EFFICIENCY OF X-RAYS

In most X-ray diffraction work, the patterns obtained are recorded for analysis on photographic plates or strips. Photographic plates, when first exposed to radiation and then developed, show results similar to those caused by exposure to visible light. The main difference is that X-rays produce an equal distribution of reduced

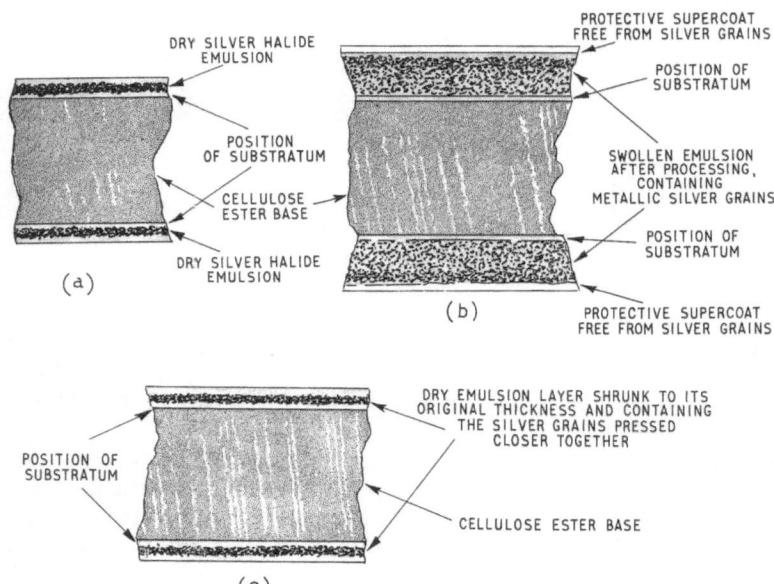

Fig. 2.7. Cross-section of an X-ray film: (a) the unexposed film; (b) the wet processed film; (c) the dried film. (Courtesy Ilford Ltd)

silver throughout the whole thickness of the sensitive layer, whereas visible light affects only the surface grains. If the thickness of the sensitive layer is increased, a greater degree of blackening can be achieved; consequently exposure times may be reduced for the same degree of blackening of film.

X-ray films of standard quality (Fig. 2.7) consist of a cellulose ester base, with a substratum and a silver halide emulsion on both sides of the plate. The effect of X-rays is to render some of the silver halide grains in the emulsion sensitive for reduction to metallic silver. This chemical reduction is accomplished in the developing solution. The developed image is occasionally visible when viewed in 'safe light'; however, the film is still light-sensitive because of the unreduced silver halide still present in the film. After careful rinsing, these remaining sensitive grains are dissolved out in the fixing bath, thus rendering the developed image visible. (The time taken to do this is known as the 'clearing time'.) Fixing should always be followed by careful washing in fresh water for at least 30 min. Finally, the film must be dried in a dust-free atmosphere.

The portions of the developed and dried plate which have been exposed to X-rays are darkened by the deposition of pure silver. This darkening D may be described quantitatively as the logarithm to the base 10 of the ratio of original intensity I_0 to the intensity I which passed through the silver deposit. Thus

$$D = \log_{10}(I_0/I)$$

This quantitative darkening is often referred to as the *optical density*. Thus, an optical density of $1\cdot0$ means that only one-tenth of the incoming light passes through the plate. A feature of X-rays is that the optical density on a photographic plate is constant for a total flux value F given by the expression

$$F = \int_0^\tau I \, d\tau$$

where I is the intensity and τ is the time of exposure.

The rules governing the optical density changes of exposed films are subject to a number of complex factors; however, for most practical problems it is sufficient to determine the response rate of a particular film by experimental methods. The film is subjected to a series of carefully graded X-ray exposures: the intensity of the radiation is kept constant, but the exposure times are successively prolonged. After a closely standardised processing sequence, the optical density of the film is measured by a densitometer. The measured optical densities are then plotted against the logarithmic values of the incident flux. Fig. 2.8 shows the variation of such

41

density in terms of the logarithm of the flux. The region lying between P and Q is the area of under exposure and that between R and S is the region of over exposure. The region of correct exposure lies between Q and R. The slope of the straight portion of the graph in the correct exposure region is called the contrast.

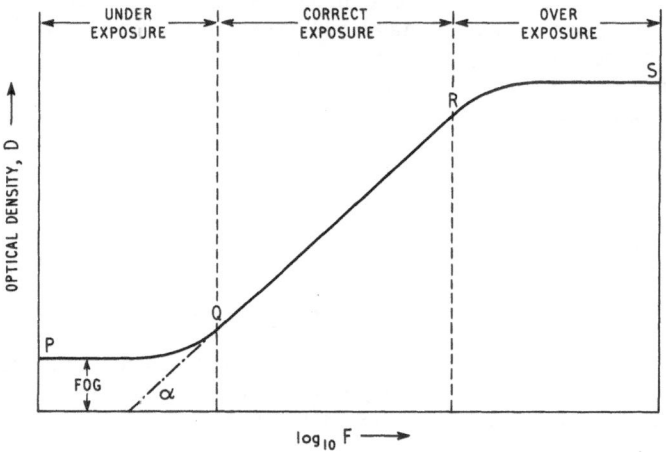

Fig. 2.8. *The exposure sensitivity of X-ray films*

In metallurgical X-ray work, it is imperative to operate within the correct range of exposures, for invariably diffraction patterns are obtained during a fixed time interval. Consequently the density variation between one diffraction spot and another provides a direct measure of the intensity variation in the diffracted beams. Intensity variations are then related to the atomic arrangement and the composition of the particular substance in question.

Manufacturers' recommendations for the correct processing of films should be followed. There are some other important points, however, to remember in order to ensure good photographic results.

Before processing begins, the temperature of the developing and fixing baths should be checked, and the temperature should be brought to 20°C. If this is not possible, developing and fixing charts must be consulted so that the recommended processing times for the prevailing ambient conditions can be used.

The level of solutions in the containers should always be kept at least 2·5 cm above the upper edge of film, and overcrowding of the tanks should be avoided. The films must be agitated vertically every minute or so, to prevent flow marks due to local concentration of free bromide and to prevent the films sticking together. If the film is immersed in the rinsing tank between each successive operation, contamination of one chemical by the other will be avoided. A solution of 2% by volume of glacial acetic acid or of potassium metabisulphite in water is recommended as rinsing agent, the effect of both being to neutralise the alkali of the developer so that it will not contaminate the fixing bath.

For a good fixing result, the clearing time of the fresh fixing bath must be recorded. As the fixing solution is used, the clearing time increases; when the clearing time is doubled, the solution should be discarded. The total fixing time is about twice the time taken to clear the film. R. W. D'Eye and E. Wait found that the one rapid fixing preparation based on ammonium thiosulphate was very convenient, as it required only a 2 min immersion time. Following the chemical processes, the film must be washed thoroughly for about 30 min in flowing water; the film, and particularly the hanger, should be drained when withdrawn for drying, as water drops may deface the semi-dry film should they drop on it.

It occasionally happens that, after about 2 h exposure and careful processing of the film, no diffraction effect appears on the plate. If it is certain that the X-ray beam entered and left the camera, then the fault invariably lies in that the developer is spent; it must be replenished or discarded completely.

2.6 X-RAY TUBES AND GENERATING EQUIPMENT

There are now many relatively inexpensive and well-tried commercial X-ray tubes and generating sets available; if this equipment is to be used to its maximum efficiency, it is important that its basic behaviour is thoroughly understood.

X-ray tubes are normally classified under the headings of cold or hot-cathode tubes. In addition, both types may be demountable or permanently sealed.

Cold cathodes are normally made of aluminium, and the source of electron flow is provided by the presence of air at about 0·01 mmHg pressure. The physical details of the X-ray generating process within the tube are not yet clear, but it is thought that the ionised atoms contained within the tube are attracted towards the cathode. The collision between the nitrogen ions in the air and the target releases a sufficient flow of electrons to maintain continuous X-ray production. The hot-cathode or Coolidge tubes are evacuated to a considerably larger extent than the cold-cathode tubes, and their source of electrons is normally provided by an electrically heated tungsten coil. In both kinds, the anode is constructed in essentially the same way. The choice of target material is governed by the scientific application intended, since only about 1 % of the available kinetic energy of the electrons in the target is transformed to X-rays and the rest is spent on heating the target; water cooling must be provided to prevent localised melting of the target.

Fig. 2.9 illustrates the principles of construction of a demountable X-ray tube. The advantage of cold-cathode tubes is that the contamination of target and windows is eliminated, and the purity of radiation is, therefore, maintained throughout the operating

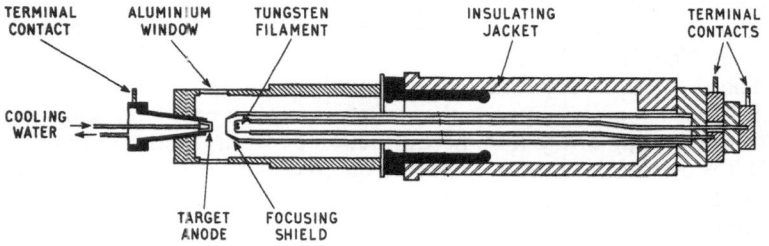

Fig. 2.9. Diagrammatic cross-section of a demountable X-ray tube

life of the tube. A greater intensity of radiation is also possible, owing to a smaller minimum focal area. For continuous service over a long period, however, hot-cathode tubes are recommended as the intensity of their radiation is more constant.

The permanently sealed X-ray tubes of modern design are invariably hot-cathode types. A sealed hot-cathode X-ray tube is shown in Fig. 2.10. The tubes and their protective shields form compact units which can be mounted in any required position. They require a very limited amount of attention throughout their operating life, and the intensity of radiation is constant over a long time. The main disadvantage is that the target material is sealed in the tube: hence, for various characteristic wavelengths, a set of interchangeable tubes must be available. There is also the problem of contamination of target and windows from the evaporating tungsten filament. Once the tube is contaminated, its operational use is fatally affected.

To overcome some of these difficulties, the modern demountable tubes were developed. Because of their demountable nature, the

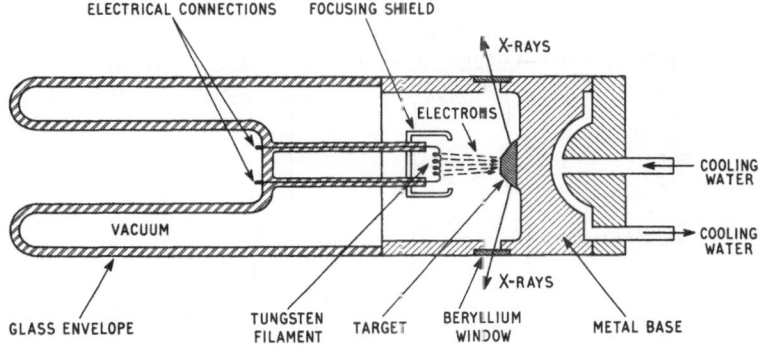

Fig. 2.10. Diagrammatic cross-section of a sealed X-ray tube

high vacuum required to operate the tube is maintained by continuous evacuation of the tube against a small controlled leakage. In a tube of this type, practically every component can be changed readily and cheaply. By changes of target, a complete range of desired wavelengths can be made available — often without even breaking the vacuum. The main disadvantage of the type, however, is that it requires about 20 min running time to provide initial evacuation. In addition, the rigid connection between the vacuum pumps and the tube reduces its manoeuvrability. It needs maintenance and attention during operation, because the variation in evacuation automatically causes variation in radiation intensity.

Hot-filament tubes of both the sealed and demountable variety are run on high-tension alternating current transformers operating from the mains. Rectification often proves unnecessary for these types of tubes, because of their self-rectifying properties. The electron beam emitted from the filament exists during the half-cycle when the filament is negative in respect to the target. In the following half-cycle, when the polarity is reversed, no electrons are emitted from the target to the filament, because the target temperature is kept well below that required for emission.

Fig. 2.11 shows a simplified wiring diagram for an X-ray generating tube. The autotransformer regulates the voltage applied

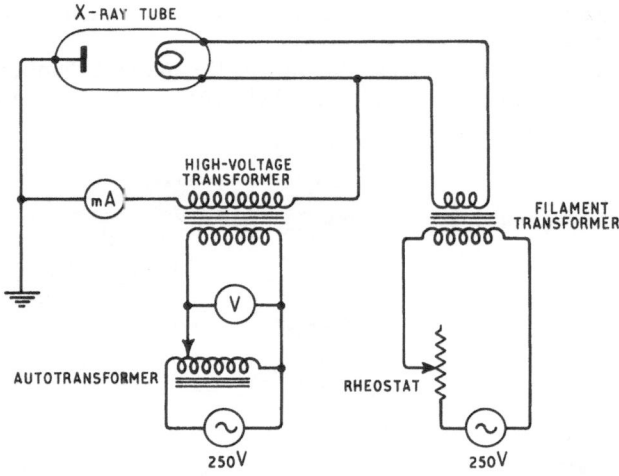

Fig. 2.11. Simplified wiring diagram for a self-rectifying X-ray tube

to the tube by controlling the applied voltage of the high-tension transformer. The voltmeter and the milliammeter are necessary for the correct adjustment of the applied voltage and the filament current. The rheostat incorporated in the circuit controls the output voltage of the filament transformer.

2.7 SAFETY PRECAUTIONS

As with other forms of radiation, X-rays are extremely dangerous to a person encountering the beam direct from its source. This is particularly true of the soft rays which are useful in crystallography. As they are less penetrating, their energies are absorbed by the human skin, causing severe burning of tissues. Several effects of X-rays on the blood supply of tissues have been studied. There is evidence of a reduction in the perfusion rate of blood through the liver after acute irradiation, together with a marked reduction in the number of red cells. It is strongly recommended that personnel engaged in routine X-ray work should have twice-yearly blood tests.

The maximum permissible dosage for the whole body is 0·5 r/week; for the hands a slightly higher value of 1·5 r/week is taken as acceptable.

To avoid radiation damage, the direct beam should first be located with a fluorescent screen attached to a probe; it should then be avoided at all times. Secondary or scattered radiation may be reduced by the use of extensive lead shields, about 1 mm thick, all round the experimental equipment. Some commercial X-ray installations and cameras have built-in safety devices which minimise the possibility of radiation exposure of the operator. For increased safety, X-ray laboratories should be monitored daily with Geiger counters or other ionisation detectors, and all personnel should wear special casettes containing X-ray sensitive film. These casettes should be processed at regular intervals, their degree of blackening being a measure of the dosage received.

Areas assigned for X-ray work should contain only the generating apparatus and the necessary peripheral equipment; they should not be used as additional working space or laboratory area. Accidental opening of an X-ray tube window may not damage the user of the equipment, but it may cause serious harm to others some distance away from the radiation source.

If accidental and unavoidable exposure is kept below the acceptable maximum level, the dangers arising from the use of X-rays are completely eliminated.

3
Crystal representation

3.1 SURVEY OF PROJECTIONS

Since the discovery of the simple law of the constancy of interfacial angles, there have been many attempts to devise methods of recording the experimental data obtained from the measurement of well-developed crystals.

Perspective drawings of crystals may show the symmetrical beauty, without necessarily yielding information on the interfacial angles. The edges which appear parallel in reality will, in the perspective representation, slope towards each other to intersect at a vanishing point. The other main types of pictorial representations, such as oblique or isometric projections, also have their limitations in that the true appearance of the crystal is sacrificed in order to gain a spatial effect in the drawing. Fig. 3.1 shows a fluorite twin drawn in perspective, together with isometric and oblique types of projections.

In some instances, particularly when the shape of the crystal is to be considered and not the relationships between its faces, the clinographic projection may usefully be employed. In the construction of a clinographic projection of a crystal of cuprous chloride, which crystallises in regular tetrahedra, the first step is to draw a clinographic axial cross. Fig. 3.2(a) shows this construction in detailed steps. The horizontal line AOA' is drawn, and BOB' is then drawn perpendicular to AOA'. The point Q is marked at a suitable distance, say 3 cm, from O. The length OP is then chosen as $\frac{1}{3}$ of OQ, i.e. 1 cm. Verticals are drawn at points P and Q, and on the vertical at P a distance PN which is $\frac{1}{3}$ of OP (i.e. $\frac{1}{3}$ cm) is marked. A straight line is drawn through ON to intersect the vertical drawn from Q at S. The line ONS is the x axis of the axial cross, with unit distance ON in that direction.

Points Z and Z' are now marked on BOB', so that $OZ = OZ' = OS$; the line ZOZ' is the z axis, and OZ is the unit distance in that direction.

The last stage in the drawing of the axial cross is the construction of the y axis. On the vertical QS, a distance QM which is $\frac{1}{3}$ of PN

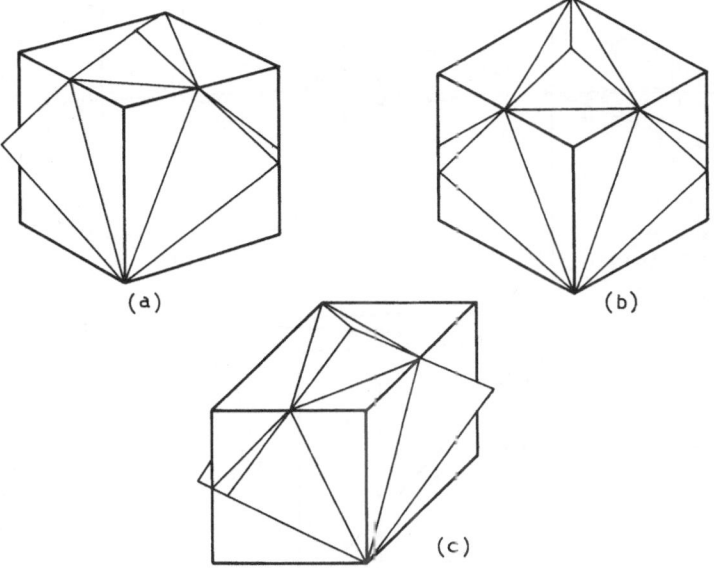

Fig. 3.1. Fluorite twin on the $[11\bar{1}]$ axis: (a) perspective representation; (b) isometric projection; (c) oblique projection

(i.e. $\frac{1}{9}$ cm) is marked; then the line OM is drawn. This line OM is now the y axis, and OM its unit distance.

Fig. 3.2(b) shows the axial cross and a cube drawn around it in the related orientation. Fig. 3.2(c) shows the tetrahedron in its correct orientation and the crystallographic axes.

Unfortunately, clinographic projections present the same difficulties as other types of projections regarding the interfacial angles of the tetrahedron. To solve the problem, it is necessary to return to basic geometry and consider the definition of the angle between two planes. This is defined either as the smallest angle

49

separating the planes; or alternatively as the angle between the normals of the planes. As the crystals are bounded by plane faces, the problem of recording true angular relationships between planes can be reduced to the problem of recording the face normals and

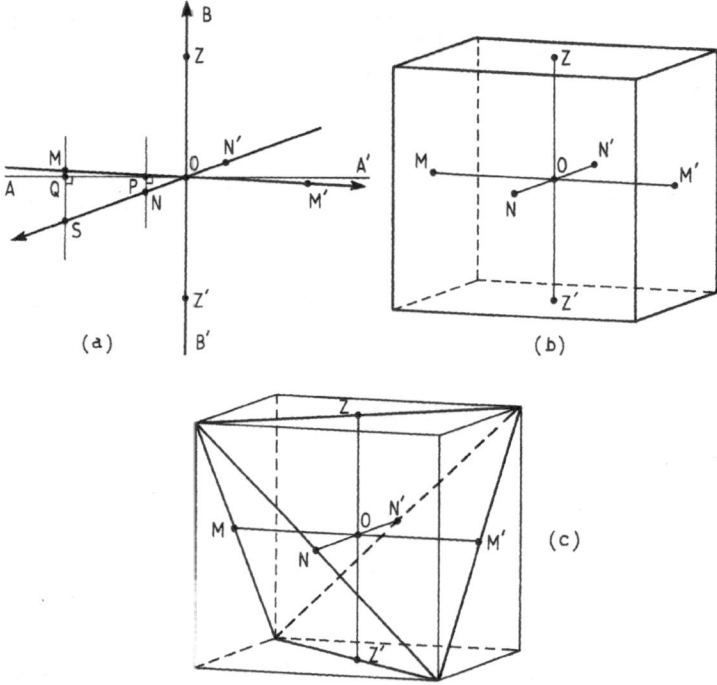

Fig. 3.2. Construction of the clinographic projection of cuprous chloride: (a) the axial cross of the cubic system; (b) the clinographic projection of a cube; (c) the tetrahedron of cuprous chloride inscribed in the cube

the angles between them. Fig. 3.3 shows a tetrahedron with its face normals *OA, OB, OC* and *OD* drawn, all originating from one point *O* within the crystal. If a sphere is now drawn around the crystal, with its centre at *O*, the normals will intersect the surface of the sphere; the relationships between these points on the sphere will be essentially similar to the relationships between the faces which the points represent. The points on the surface of the sphere are called the *face poles*. The descriptive angular relationships

50

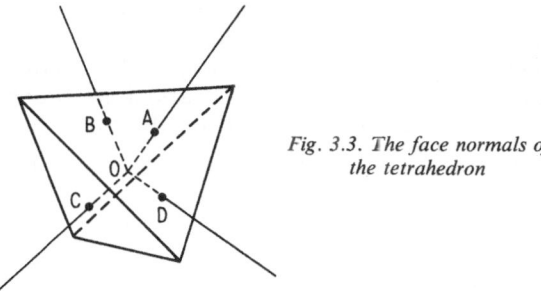

Fig. 3.3. *The face normals of the tetrahedron*

observable in a crystal are thus represented on a sphere; the construction is called the *spherical projection*.

The spherical projection, though useful, is still three-dimensional and it is therefore difficult to use in the two-dimensional plane of the drawing paper. The problem of reducing the three-dimensional sphere to the plane of the paper is essentially a navigational and cartographic concept, the principles involved being similar to those used in the representation of the continents and seas of the globe on two-dimensional maps.

Consider again the spherical projection of the tetrahedron and its face poles. If plane *m* is inserted in the equatorial position (Fig. 3.4), it may be used as the plane of projection. In order to project the face poles located on the upper hemisphere on to the equatorial plane, a line is drawn through each face pole and the pole of the lower

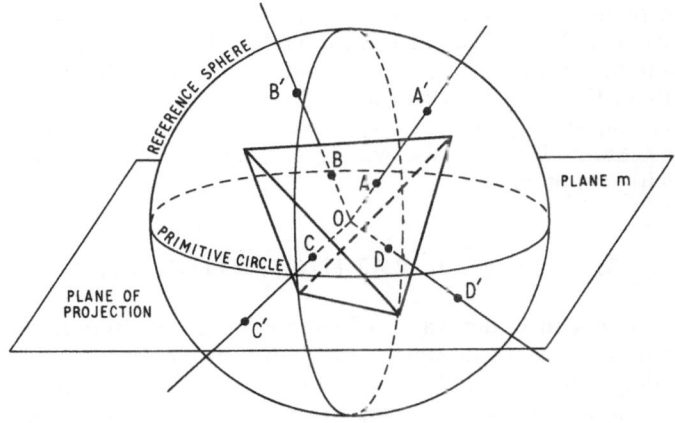

Fig. 3.4. *The spherical projection of the tetrahedron*

51

hemisphere (pole *S* in Fig. 3.5). The point where the projecting ray intersects plane *m* will then be the stereographic projection of the face pole in question. Face poles projected from the pole of the lower hemisphere are traditionally marked in projection by a black dot. If the face pole is located on the lower hemisphere, it

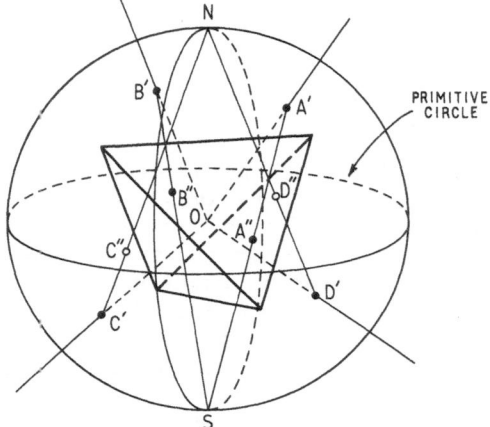

Fig. 3.5. The stereographic projection of the face poles

is projected from the pole of the upper hemisphere and the projection is indicated by an open circle.

The plane of projection cuts the sphere in a great circle known as the *primitive*. If the face poles are joined to the appropriate pole of projection, the projection of the points will always lie within the primitive. For ease of construction and versatility of application, the stereographic projection occupies a unique place in crystallography. Examples of its use will be constantly encountered in the following chapters.

3.2 STEREOGRAPHIC PROJECTION

If the practicality and value of stereographic projections (which incidentally are encountered in many disciplines other than crystallography or metallurgy) are to be appreciated, some of their properties should be examined. There are two basic properties which distinguish stereographic projections from all other methods.

The first is that the projection of a circle drawn on the sphere projects on to the plane of projection as a circle. The second is that the angular truth is preserved in the projection.

The first principle simplifies the drawing of projections. Great circles of any orientation will project within the primitive as arcs of circles, degenerating into a straight line through the centre of the primitive as the great circle reaches a vertical position. Small circles project as circles lying within the primitive. The basic stereographic constructions are introduced by the following series of simple examples.

3.2.1 CONSTRUCTION OF A GREAT CIRCLE, IN PROJECTION, THROUGH TWO GIVEN POLES X AND Y IN THE EQUATORIAL PLANE

In Fig. 3.6, the two poles X and Y are marked; they are, in fact, projections on the equatorial plane. O is the centre of the primitive. The line XOA is drawn to intersect the primitive at A, and a line

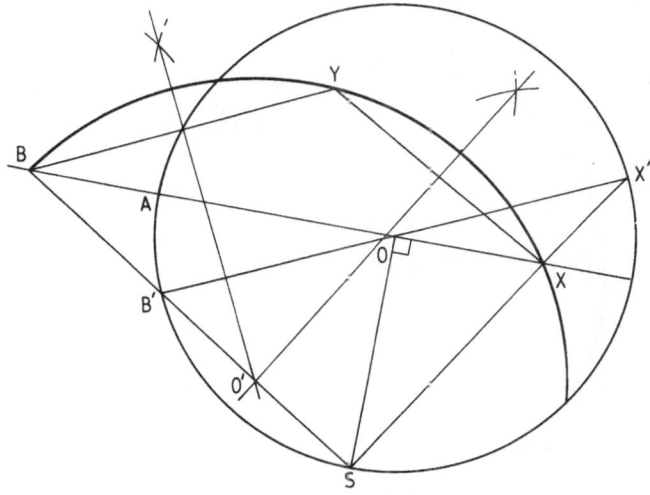

Fig. 3.6. Construction of a great circle through two given poles

constructed from O perpendicular to XOA intersects the primitive at S. SX is projected on to the primitive at point X', and a diameter is drawn from X' through O to intersect the primitive at B'. The

53

stereographic projection of B' is located at B, which is found by projecting SB' and XOA until they intersect. To find the centre of the great circle through the three points X, Y and B, the points are first connected by chords. The perpendicular bisectors of these chords then meet at O', which marks the desired centre.

A detailed examination of this construction shows that the primitive circle has in fact been used as the contour of the sphere of projection, and the line XOA has been taken as the trace of the equatorial plane. This simplification of geometrical construction is used in all the following examples.

3.2.2 CONSTRUCTION OF A GREAT CIRCLE, IN PROJECTION, AT 90° TO A GIVEN POLE X IN THE EQUATORIAL PLANE

In Fig. 3.7, the pole X is marked. The diameter OX (the trace of the equatorial plane) is drawn, and a line perpendicular to OX is dropped from O. This perpendicular intersects the primitive (the trace of the sphere of projection) at point P (the pole of projection).

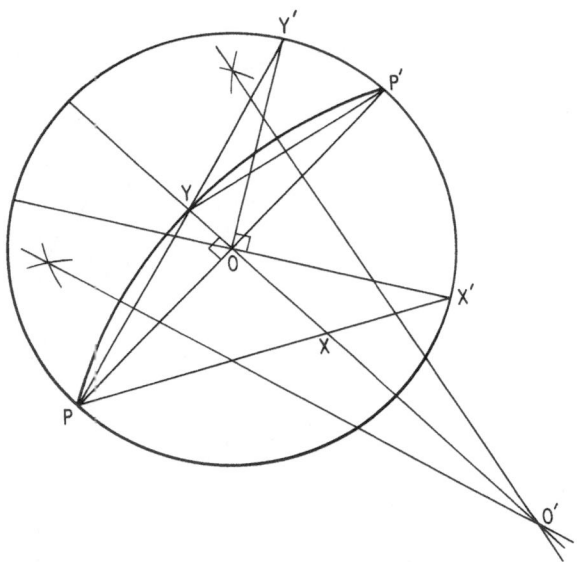

Fig. 3.7. Construction of a great circle at 90° to a given pole

Projection of PX on to the primitive gives the point X'. The line OX' and the line perpendicular to OX' through the centre O are then drawn. This perpendicular intersects the primitive at Y'. The point Y' is projected back on to the trace of the equatorial plane (i.e. the line $Y'P$ is drawn); at the intersection of lines $Y'P$ and OX is the point Y, which is at $90°$ to X and is located on the diameter OX. The point P, and the point P' lying diametrically opposite to P, are also points on the same great circle. As three points all lying on the same great circle have now been found, the great circle may be drawn as in the preceding example.

3.2.3 MEASUREMENT OF THE ANGLE BETWEEN TWO GIVEN GREAT CIRCLES A AND B

In Fig. 3.8, two great circles A and B are shown intersecting at point P. A great circle at $90°$ to P is drawn in a similar manner to

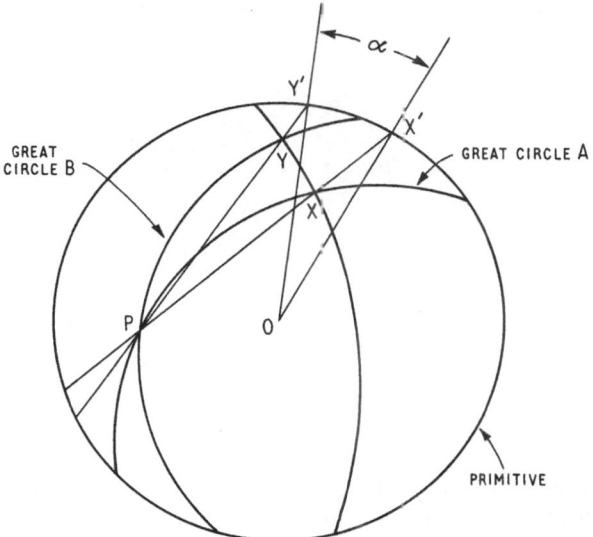

Fig. 3.8. Construction for the measurement of the angle between two given great circles

that described in Section 3.2.2. This great circle intersects the given two great circles A and B at X and Y respectively, both of which poles are at $90°$ to P. Points X and Y are projected from P on to the primitive, which results in points X' and Y'. The angle α between the great circles A and B is given by angle $X'OY'$, as indicated in the diagram.

3.2.4 FINDING A THIRD POLE WHICH IS AT TWO GIVEN ANGLES α AND β TO TWO GIVEN POLES X AND Y

To simplify construction, poles X and Y in Fig. 3.9 are assumed to lie on the primitive. The position of the third pole P is fixed by angles α and β measured from X and Y respectively. The desired pole P then must lie on a small circle of radius α around X, and

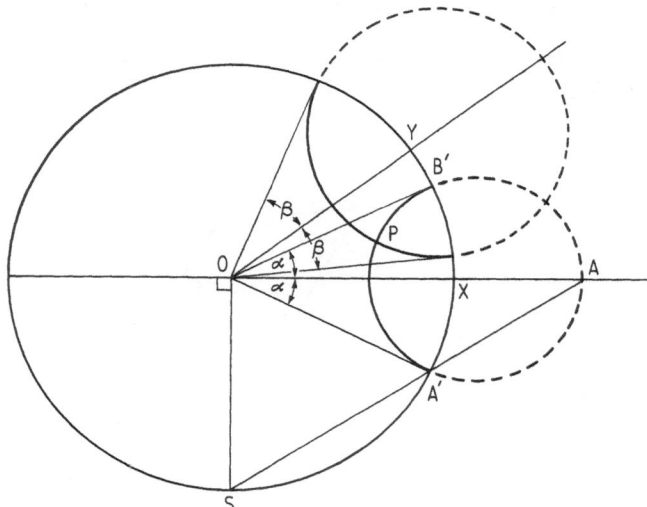

Fig. 3.9. Construction of a pole lying at two given angles to two given poles

simultaneously on a small circle of radius β around Y; thus the pole P must lie at the intersection of these two small circles. To construct the two small circles, the diameter XO is first drawn, and then a perpendicular OS. An angle α at O is measured as shown in Fig. 3.9. The rays drawn from O intersect the primitive at points A' and B'. The point A' is projected from S on to the line OX, giving point A. The three points A', A and B' will then lie on a small circle with radius α around X. To find the centre of this small circle (which is *not* point X), it is only necessary to draw chords and their perpendicular bisectors. The intersection of the bisector will mark the centre of the small circle. The construction is now repeated for pole Y using angle β. The intersection of the two small circles produces the required pole.

To reduce the tedium of an excessive amount of graphical work, there are many auxiliary devices available from commercial sources. The most useful of these devices are the stereographic or Wulff net

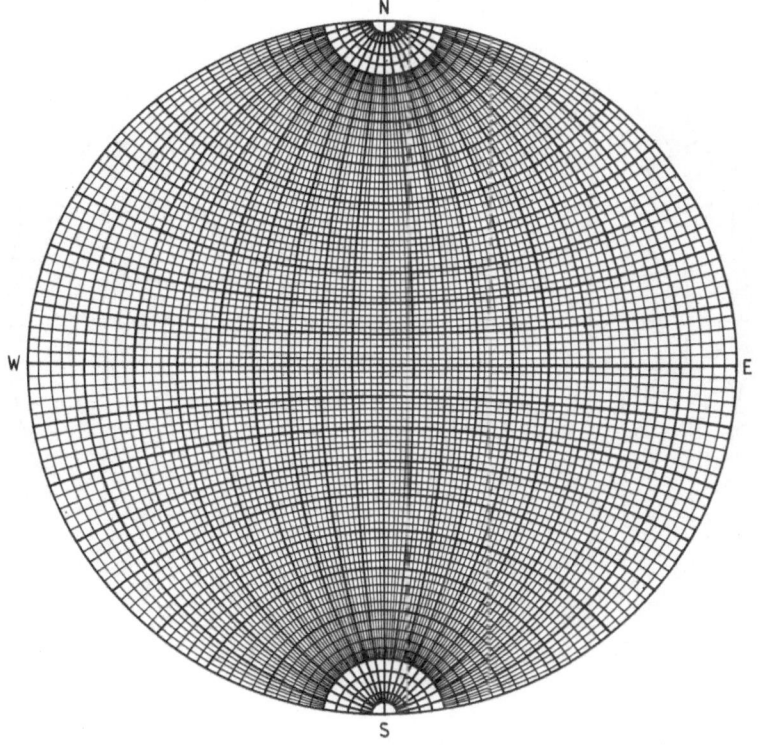

Fig. 3.10. The stereographic or Wulff net

(Fig. 3.10) and the polar net (Fig. 3.11). The Wulff net is made up of a series of small and great circles, usually drawn at 2° intervals and normally produced on transparent paper. (Nets can be obtained from The Institute of Physics and The Physical Society, London.)

Fig. 3.12(a) shows how a stereographic net can be used to draw a great circle through two given poles. The Wulff net is superimposed upon the projection so that the centre of the primitive circle coincides with the centre of the Wulff net. By rotation of the Wulff

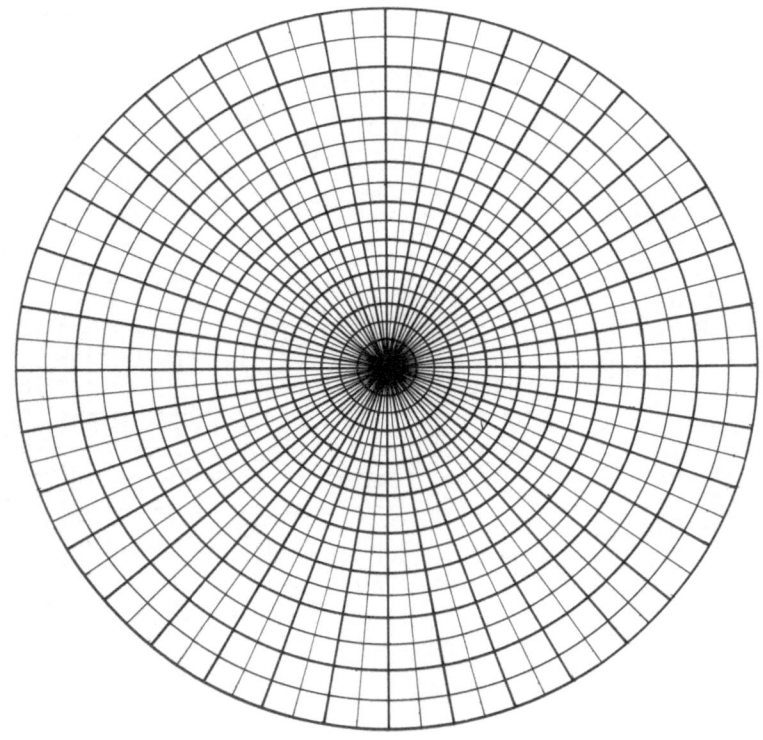

Fig. 3.11. The polar net

net around its centre, a great circle may be found that passes through both poles. The great circle may then be copied on to the paper of the projection by pricking with a pin through the Wulff net. The other diagrams in Fig. 3.12 demonstrate other uses of the Wulff net.

58

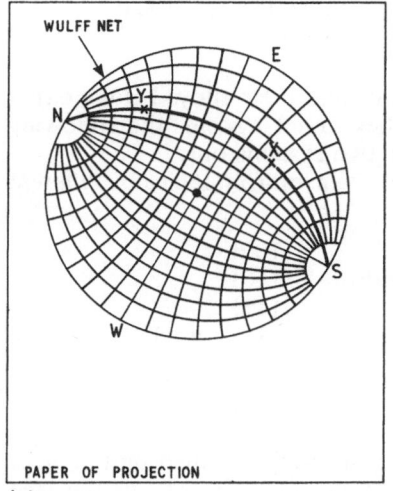

(a)

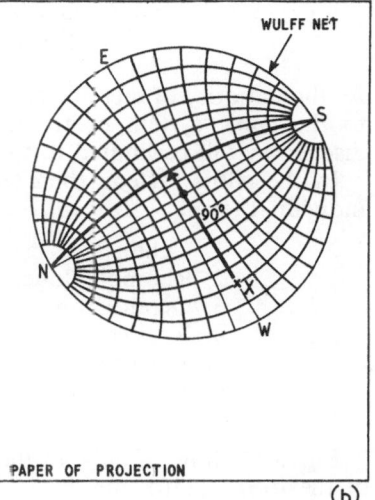

(b)

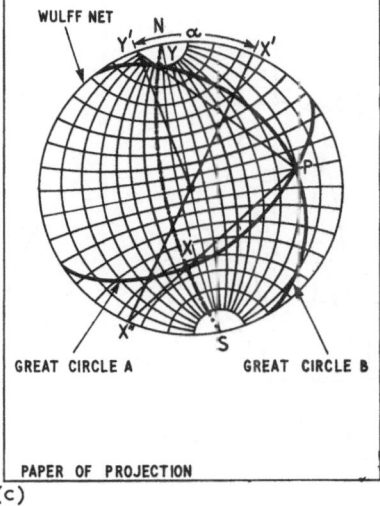

(c)

Fig. 3.12. Applications of the Wulff net: (a) to drawing a great circle through two given poles; (b) to drawing a great circle at 90° to a given pole; (c) to measuring the angle between two great circles

3.3 STEREOGRAMS OF CUBIC AND HEXAGONAL SYSTEMS

As the majority of metallic elements and their alloys belong either to the cubic system or to the hexagonal system, the following discussion will be limited to these two systems only.

In the cubic system, there are seven morphologically different individual forms:

1.	Cube	$\{100\}$	
2.	Rhombic dodecahedron	$\{110\}$	
3.	Octahedron	$\{111\}$	
4.	Icositetrahedra	$\{h11\}$	$h>1$
5.	Trisoctahedra	$\{hh1\}$	$h>1$
6.	Tetrahexahedra	$\{hk0\}$	
7.	Hexoctahedra	$\{hk1\}$	

Fig. 3.13 shows a cube of the form $\{100\}$ modified by the faces of the forms $\{110\}$ and $\{111\}$. The diagram also shows the face normals of some selected planes. The poles of planes, which are so arranged that the edges formed by them are parallel, also lie on the same great circle. Such a set of faces, whose mutual intersections

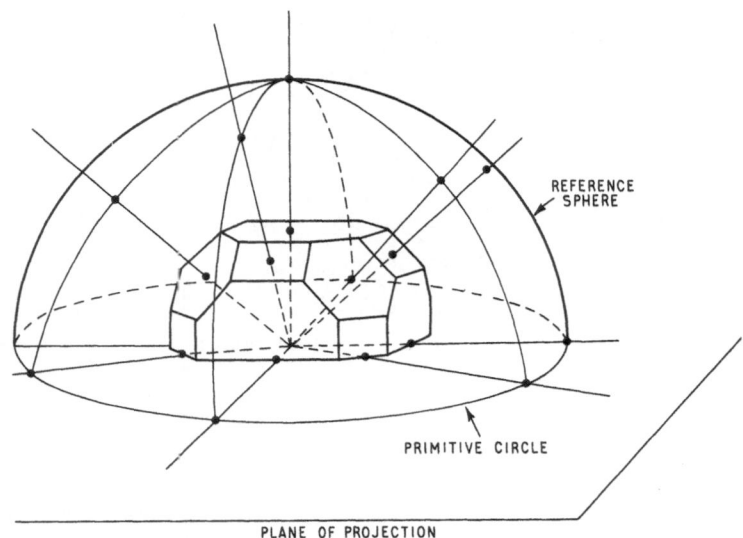

Fig. 3.13. The face normals of a modified cube

are all parallel, constitute a zone. The common direction of the edges is known as the zone axis. With this simple zone relationship in mind, the construction of the stereogram can be started. It is conventionally agreed that the primitive circle should be drawn with a diameter of 5 in (but for convenience a smaller primitive circle is used in this book).

By inspection of Fig. 3.13, the faces of the crystal may be grouped into nine zones as follows:

Zone	Indices of faces
a	(100), (110), (010), ($\bar{1}$10), ($\bar{1}$00), ($\bar{1}\bar{1}$0), (0$\bar{1}$0), (1$\bar{1}$0)
b	(100), (101), (001), ($\bar{1}$01), ($\bar{1}$00)
c	(010), (011), (001), (0$\bar{1}$1), (0$\bar{1}$0)
d	(110), (111), (001), ($\bar{1}\bar{1}$1), ($\bar{1}\bar{1}$0)
e	(1$\bar{1}$0), (1$\bar{1}$1), (001), ($\bar{1}$11), ($\bar{1}$10)
f	(100), (1$\bar{1}$1), (0$\bar{1}$1), ($\bar{1}\bar{1}$1), ($\bar{1}$00)
g	(100), (111), (011), ($\bar{1}$11), ($\bar{1}$00)
h	(010), (111), (101), (1$\bar{1}$1), (0$\bar{1}$0)
i	(010), ($\bar{1}$11), ($\bar{1}$01), ($\bar{1}\bar{1}$1), (0$\bar{1}$0)

Zone a may be readily plotted on the primitive, because all the poles lying on the equatorial circle also lie on the primitive circle in the plane of projection. Inspection of Fig. 3.13 also shows that, in the cubic system, faces of the form $\{110\}$ lie at 45° to the faces of the form $\{100\}$. Zones b, c, d and e appear as straight lines connecting the poles lying in the same zones. Fig. 3.14(a) illustrates the construction so far completed.

The positions of the poles (101), (011), ($\bar{1}$01) and (0$\bar{1}$1) are now marked on the stereogram, as shown in Fig. 3.14(b). Again these projected poles lie at an angle of 45° between pole (001) as centre and the poles (100), (010), ($\bar{1}$00) and (0$\bar{1}$0) respectively. These positions can be measured with the help of a Wulff net. Finally, the positions of the form $\{111\}$ should be fixed; they may be located by direct measurements of interfacial angles, although this is not really necessary. For example, it should be noticed that the pole ($\bar{1}\bar{1}$1) lies in both zone f and zone d. Zone f may easily be drawn using the Wulff net, as poles (100), (0$\bar{1}$1) and ($\bar{1}$00) are already shown in the diagram. The pole ($\bar{1}\bar{1}$1) will then lie on the great circle drawn through these poles. Since the pole ($\bar{1}\bar{1}$1) also belongs to zone d, the projected pole must lie at the intersection of the two zones. The remaining three poles can be found by similar reasoning. The complete stereogram of the modified cube is shown in Fig. 3.14(c).

The stereogram of Fig. 3.14(c) contains some of the important planes of the cubic system, but it would be much more useful if it included some higher order pole projections. In Fig. 3.15, some of the poles of the forms $\{h11\}$, $\{hh1\}$, $\{hk0\}$ and $\{hk1\}$ are also included. Fig. 3.16 and Fig. 3.17 show the (110) and the (111) stereographic projections of the cubic system.

At this stage, it is useful to drop the idea of rigid adherence to crystal morphology. In Chapter 1, it was shown that the actual

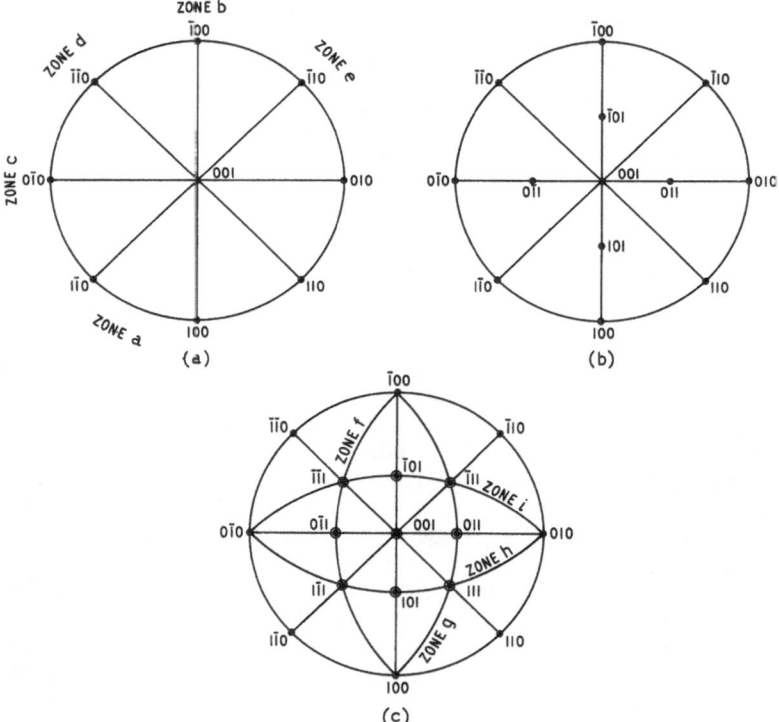

Fig. 3.14. The construction of a stereogram of the cubic crystal shown in Fig. 3.13.

faces or cleavage planes of crystals are *ipso facto* atomic planes. Therefore, the projected poles of the stereogram can be directly related to the atomic planes whose Miller indices are similar to those of the actual crystal faces. It is clear that the three stereograms of

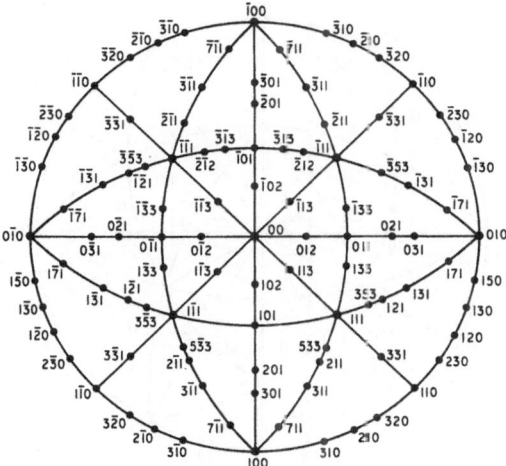

Fig. 3.15. The standard (001) projection of the cubic system

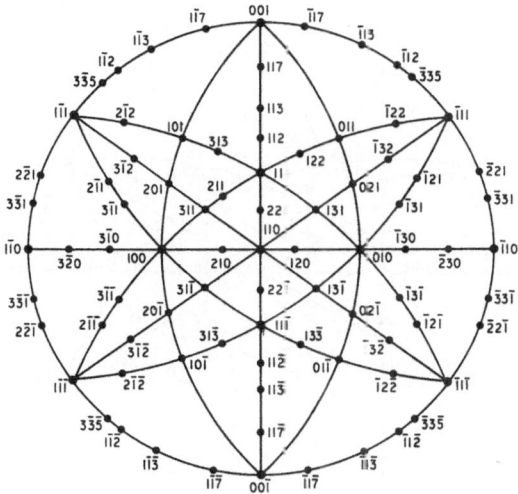

Fig. 3.16. The standard (110) projection of the cubic system

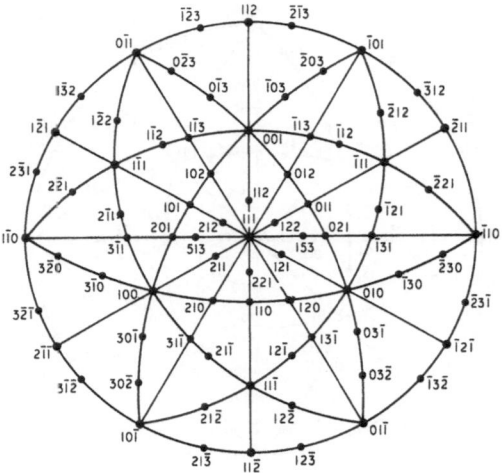

Fig. 3.17. The standard (111) projection of the cubic system

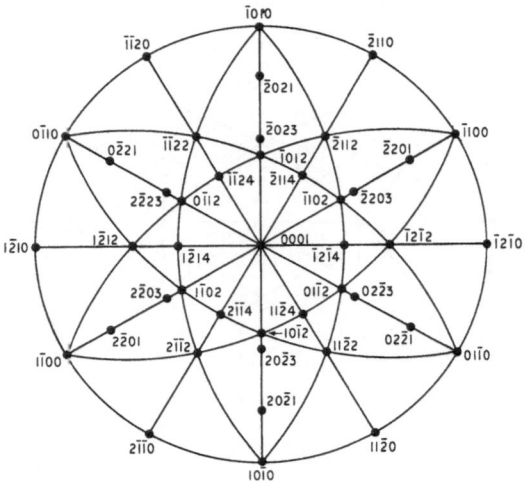

Fig. 3.18. The (0001) projection of magnesium

Fig. 3.15, Fig. 3.16 and Fig. 3.17 show the angular relationships between a number of atomic planes, as well as between possible faces of the cubic system.

Another useful property of the stereogram of the cubic system is that it is a true representation of all substances which crystallise in the cubic system: changing the unit cell dimension will not alter any angular relationships within the cell. This unfortunately does not apply to any other crystal systems. Because of the variations in axial ratios in the hexagonal system, it is not possible to construct a standard stereogram for the whole system. Fig. 3.18 shows the standard (0001) projection for magnesium (axial ratio $c/a = 1.623$). The stereogram for magnesium approaches the ideal arrangement of poles, because its axial ratio is very close to the ideal close-packed ratio of 1.633.

3.4 SOME MATHEMATICAL RELATIONSHIPS

In some instances, particularly when greater accuracy is required than that obtained geometrically, certain mathematical relationships may be of use. The following sections give the more important calculations required.

3.4.1 THE ANGLE BETWEEN THE TWO FACE NORMALS

The angle θ between the normals to the planes (hkl) and (HKL) in the cubic system is given by the equation:

$$\cos \theta = \frac{hH + kK + lL}{[(h^2 + k^2 + l^2)(H^2 + K^2 + L^2)]^{\frac{1}{2}}} \tag{3.1}$$

The interplanar angle θ in the hexagonal system between the planes $(hkil)$ and $(HKIL)$ is calculated from the formula:

$$\cos \theta = \frac{hH + kK + \frac{1}{2}(kH + hK) + \frac{3}{4}(a/c)^2 lL}{\{[H^2 + HK + K^2 + \frac{3}{4}(a/c)^2 L^2][h^2 + hk + k^2 + \frac{3}{4}(a/c)^2 l^2]\}^{\frac{1}{2}}} \tag{3.2}$$

As the system becomes less symmetrical, the equation grows more and more complex. For instance, in the triclinic system, the equation of interfacial angles would include the six indices of the two planes, three values of the axial angles and the three unit cell parameters. The greatest simplification could be effected by

65

calculating the slope of the faces relative to one of the reference coordinate axes, and then solving the spherical triangle to give the required result.

3.4.2 THE ZONE SYMBOL

A zone is defined as a set of planes with mutually parallel inter-sections, and this common direction of the edges is called the *zone axis*. The indices of this general direction $[UVW]$ are called the *zone symbol*. Let the indices of the two faces constituting a zone be (hkl) and (HKL) then:

$$\left.\begin{array}{l} U = kL - lK \\ V = lH - hL \\ W = hK - kH \end{array}\right\} \tag{3.3}$$

For the numerical evaluation of the zone symbol equations, some simple methods have been devised. To demonstrate the principles, let the two faces of the zone be $(10\bar{1})$ and (121). Each index is written twice, the second index below the first, thus:

$$1 \ 0 \ \bar{1} \ 1 \ 0 \ \bar{1}$$
$$1 \ 2 \ 1 \ 1 \ 2 \ 1$$

The first and last columns are deleted, and the figures are then cross multiplied, the product of a pair joined by a dotted line being subtracted from the product of a pair joined by a solid line.

This gives:

$$0 \times 1 - (-1) \times 2 = 2$$
$$(-1) \times 1 - 1 \times 1 = -2$$
$$1 \times 2 - 0 \times 1 \quad = 2$$

Since factorising does not change the direction of vectors, it is possible to divide through by two to give the zone symbol $[1\bar{1}1]$ in the form of least integers.

66

3.4.3 THE ZONE LAW

Converse problems may also arise in stereographic projections. For instance, the problem may be to establish whether a certain face with indices (hkl) belongs to the zone $[UVW]$. The Weiss zone law states that the sum of the multiples of the indices is zero, thus:

$$Uh + Vk + Wl = 0$$

In Section 3.4.2, it was established that the two faces $(10\bar{1})$ and (121) form the zone $[1\bar{1}1]$. The indices of all other faces belonging to the same zone should satisfy the zonal law. Therefore, for all faces (hkl) in this zone:

$$h - k + l = 0 \tag{3.4}$$

It follows from the zone law that, if the poles $(h_1 k_1 l_1)$ and $(h_2 k_2 l_2)$ lie in the same zone and their projections lie on a great circle, then the indices $(h_3 k_3 l_3)$ of a pole lying on the same great circle and located between the two given poles are

$$\left. \begin{aligned} h_3 &= h_1 + h_2 \\ k_3 &= k_1 + k_2 \\ l_3 &= l_1 + l_2 \end{aligned} \right\} \tag{3.5}$$

For instance, if the poles (110) and (010) lie on a great circle, the pole located between them is the sum of the individual indices, i.e. (120).

This result can be checked by means of the zone law. The zone symbol of faces (110) and (010) is given by:

$$
\begin{array}{c|cccc|c}
1 & 1 & 0 & 1 & 1 & 0 \\
0 & 1 & 0 & 0 & 1 & 0
\end{array}
$$

Thus, the zone symbol is $[001]$. If the face (120) is in the same zone, then

$$0 \times 1 + 0 \times 2 + 1 \times 0 = 0$$

Since this is so, it confirms that the face (120) is tautozonal with the faces (110) and (010).

4

X-ray diffraction by crystalline matter

4.1 DIFFRACTION BY LATTICE PLANES

As with visible light or radio waves, X-rays may be regarded as electromagnetic disturbances propagating from a point of origin. If an electron happens to be in the path of propagation, it will resonate with the disturbance which passes it. This forced oscillation involves the electron in some changes of velocity, and the accelerating or decelerating electron itself becomes a source of disturbance; the electron is said to scatter the incident disturbance. Since the resonant vibration, by definition, retains the phase and frequency of the incident vibration, the scattered disturbance is in phase with the incident disturbance and is of the same wavelength.

Light waves are diffracted by optical gratings. An optical grating consists of evenly spaced transparent strips separated by equal parallel opaque strips. Consider now a space lattice, with atoms located at the lattice points. Each of these regularly spaced atoms contains a number of electrons; and the atoms are separated by space containing non-scattering media. The similarity of this arrangement to an optical grating is obvious, almost the only difference being the relative sizes. The longer wavelength of visible light requires a grating of approximately 5,000 transparent strips per centimetre; whereas X-rays, having wavelengths between about 0.5×10^{-8} cm and 3×10^{-8} cm, require a grating of approximately 12×10^8 'transparent' strips per centimetre.

The famous experiment of von Laue, in 1912, was the first to demonstrate the diffraction of X-rays by a crystal. In his experiment, a narrow beam of heterochromatic X-rays was passed through a crystal of zinc blende and was allowed to fall upon a photographic plate. The developed plate showed a pattern of spots around a central intense spot caused by the undeviated X-ray beam. Examples of various Laue-type photographs are shown in Fig. 4.1.

Fig. 4.2 helps to illustrate how this diffraction can occur. Assume that a beam of parallel monochromatic X-rays (bounded by the rays $X'X$ and $Y'Y$) falls upon an atomic plane XY. The atoms of this plane XY will then act as sources of scattered radiation. The scattered radiation of the individual atoms builds up a reflected wave in accordance with the Huygen construction. If XZ is drawn perpendicular to $Y'Y$, then XZ represents the wavefront of the incident radiation (i.e. all points along XZ are in phase when the ray $X'X$ reaches the plane XY). Thus, during the time taken by the disturbance at Z to reach point Y, the point X will emit scattered radiation of the same velocity as the incident radiation. Hence the circle of radius ZY and centre X represents the position, at the instant when the wavefront represented by XZ reaches Y, of the scattered disturbance emitted from X when this wavefront reached X. The wavefront of the scattered radiation from X is then the tangent $X''Y$ of this circle. Just as the wavefront of the incident radiation is perpendicular to its direction of propagation, so the line XX'', perpendicular to the wavefront $X''Y$, represents the diffraction direction.

It follows from geometrical considerations that, if $X''Y$ is the diffracted wavefront, the scattering sources must be evenly spaced. The Huygen construction may be repeated in the plane normal to that of Fig. 4.2, the plane in which the perpendicular wavefronts lie being called the wavesurface.

The atomic plane XY diffracts only a small proportion of the incoming energy; the transmitted wave passes on, encountering other atomic planes where diffraction again occurs. If the new diffracted beam happens to be in phase with the previously diffracted beam, their individual energies are additive. In general, atomic planes must obey certain geometrical restrictions if the diffracted beams are to reinforce each other.

Fig. 4.3 shows a section through a crystal, two parallel atomic planes being marked p and q. The atomic planes are separated by a distance d, and the glancing angle of the incident X-rays is θ. The line AD is drawn perpendicular to the incoming beam, as in the Huygen construction. For constructive diffraction to occur, the rays DD'' and BB'' must be in phase, and this can only occur if the distance ABC is an integral multiple of the wavelength λ. If BD is

69

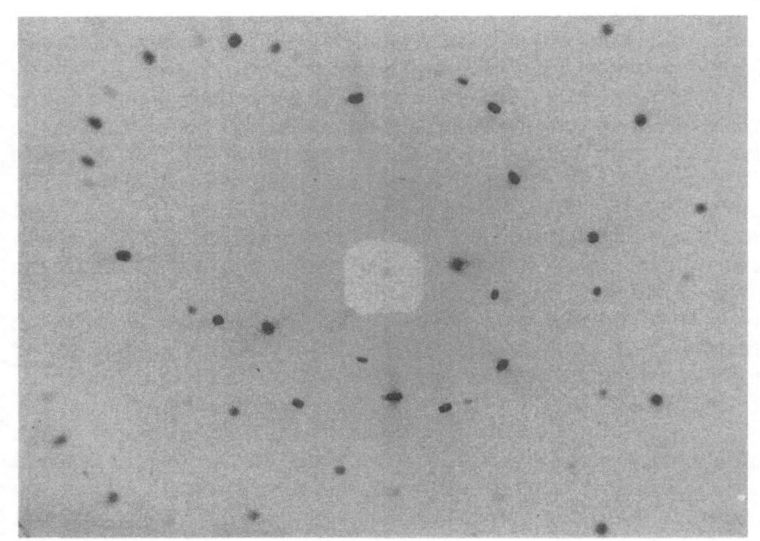

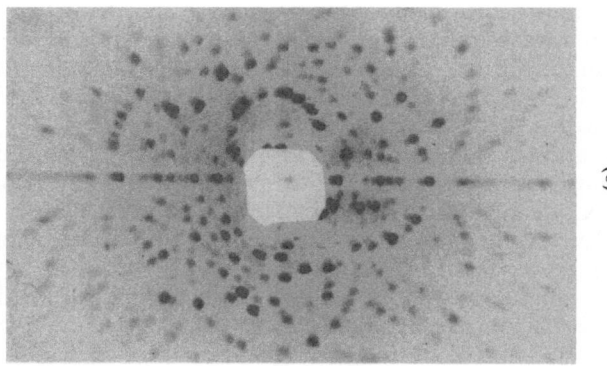

(a)

(b)

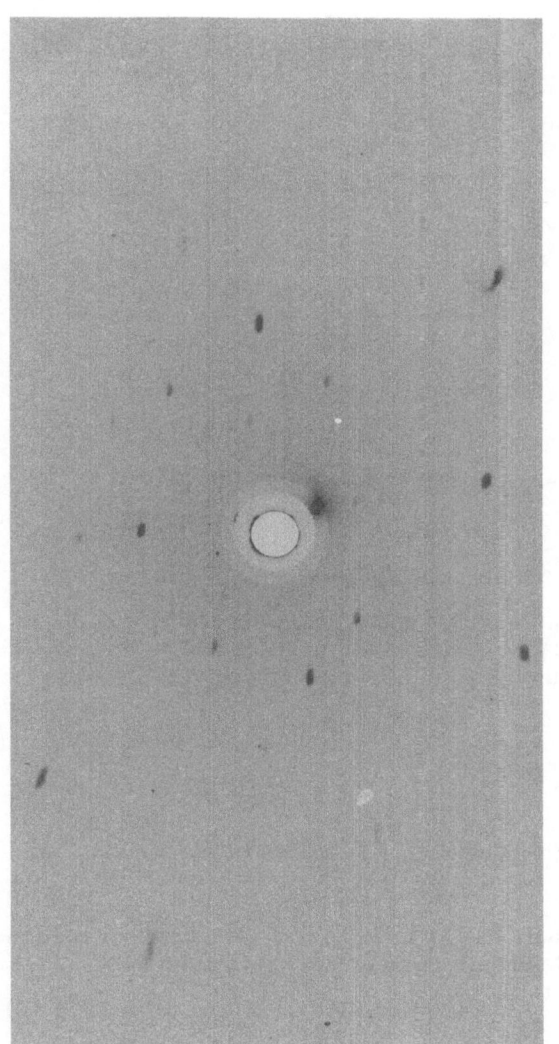

(c)

Fig. 4.1. Examples of Laue photographs: (a) transmission photograph of single crystal thiosulphate obtained with unfiltered copper radiation at 35 kV and 13 mA, exposure time 30 min, film-to-specimen distance 3 cm; (b) transmission photograph of single crystal aluminium obtained with unfiltered copper radiation at 30 kV and 13 mA, exposure time 25 min, film-to-specimen distance 3 cm; (c) back-reflection photograph of single crystal aluminium obtained with unfiltered copper radiation at 35 kV and 15 mA, exposure time 40 min, film-to-specimen distance 3 cm

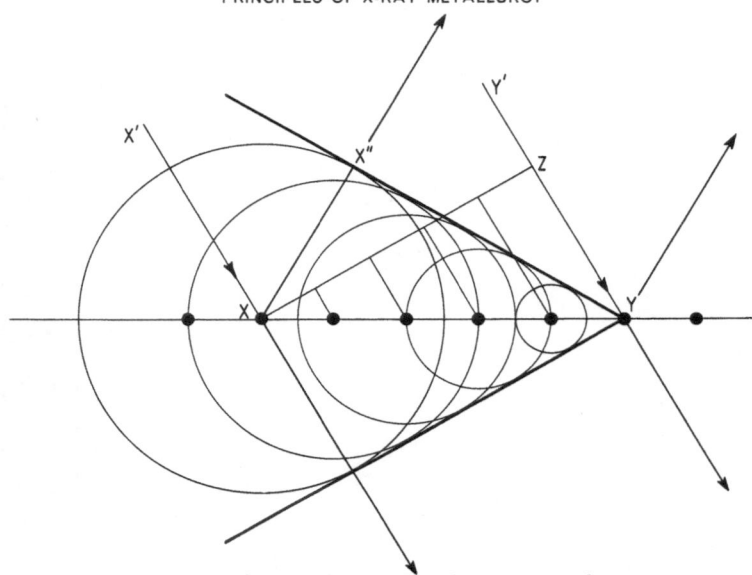

Fig. 4.2. Reflection of a wavefront by an atomic plane

drawn perpendicular to the planes *p* and *q*, then angles *ADB* and *BDC* are both equal to the glancing angle θ. The condition for reinforcement is thus

$$AB + BC = n\lambda \qquad (4.1)$$

where *n* is an integer. But

$$AB = BC = d \sin \theta \qquad (4.2)$$

Hence the total path difference $AB + BC$ can be written as:

$$AB + BC = 2d \sin \theta \qquad (4.3)$$

Substitution of Equation 4.3 into Equation 4.1 gives the condition for constructive interference of the diffracted rays as:

$$n\lambda = 2d \sin \theta \qquad (4.4)$$

This equation was first formulated by W. L. Bragg, and is known as the Bragg law. The value of *n*, as in optics, determines the order of reflection; the upper limit is given by the relationship

$$\sin \theta = \frac{n\lambda}{2d} < 1 \qquad (4.5)$$

72

The fundamental difference between the diffraction of X-rays by crystalline matter and the reflection of visible light is that the X-ray diffraction effect builds up within the whole body of the crystal (and its interpretation therefore provides information on the complete crystal); the interpretation of light reflection, on the other hand, only indicates the conditions on the surface of the reflecting matter. A second difference between the two phenomena is that light waves reflect at any angle of incidence, whereas X-ray diffraction is rigidly governed by the Bragg equation.

The experiment of von Laue on X-ray diffraction proved two points simultaneously. It was already known from the study of

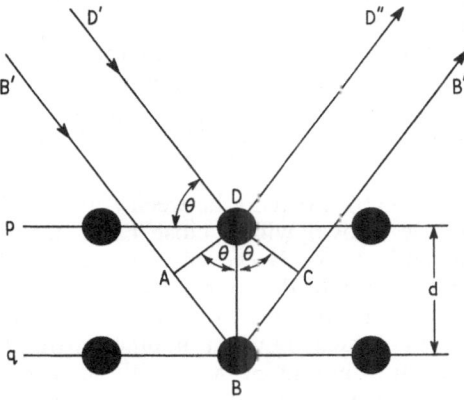

Fig. 4.3. Reflection of rays by two atomic planes

optics that diffraction could only occur in a wave-like disturbance encountering some regularly spaced obstacles. Thus, the success of von Laue showed beyond doubt the wave nature of X-rays, and that the crystalline state means an orderly arrangement of scattering media in three dimensions.

4.2 INTENSITY MODIFICATIONS OF DIFFRACTED X-RAYS

The Bragg law, as developed in Section 4.1, yields the strict geometrical relationship between the wavelength of incoming X-rays, the glancing angle and the lattice spacing. However, the Bragg law contains no information on the intensity of the diffracted beam.

73

The intensity of the diffracted beam must basically depend upon the kind of atom which causes its scattering. Since the scattering of the X-rays occurs in the electron shells, it is the number of electrons which prescribes the scattering efficiency of the atom. This scattering efficiency of an individual atom is called its *atomic scattering factor* and is denoted by f_0. If the atom is assumed to be a point, occupying no space whatsoever, the atomic scattering factor is equal to the atomic number Z. However, if the wavelengths and interatomic distances are considered in real dimensions, the size of the atoms involved may no longer be assumed to be zero. The scattered waves originating from electrons located in various positions around the atom must have some minor phase differences, resulting in a strengthening of intensity in some directions and a weakening in others. It has been found that f_0 is a function of $\sin(\theta/\lambda)$, and may be expressed by the equation

$$f_0 = \int_0^\infty \frac{U(r) \sin\left[4\pi r \sin(\theta/\lambda)\right]}{4\pi r \sin(\theta/\lambda)} \, dr \qquad (4.6)$$

where $U(r)$ is the electron density between r and $r+dr$.

Values of f_0 are known within close limits from the work of Pauling, Fermi and others. However, these scattering factors are derived by theoretical considerations, and the atoms are assumed to be at absolute zero temperature. At other temperatures, the atoms oscillate in a near random manner around their lattice position. To a train of impinging X-rays, these atoms appear to be displaced from their theoretically true position; and these displacements cause small differences in phase. Hence, the scattering factor for normal temperatures must be modified by a factor equal to

$$\exp\left[-K\left(\frac{\sin\theta}{\lambda}\right)^2\right]$$

where K is a constant related to the amplitude of the temperature-induced atomic vibration. The atomic scattering factor f may thus be calculated from the equation:

$$f = f_0 \exp\left[-K\left(\frac{\sin\theta}{\lambda}\right)^2\right] \qquad (4.7)$$

The variation of the atomic scattering factor of some elements with $(\sin\theta)/\lambda$ is shown in Fig. 4.4 (λ is multiplied by 10^8 to convert from Ångstroms to centimetres).

74

The amplititudes and phases of waves scattered by the atoms of a unit cell may be added vectorially. Suppose that the unit cell of a substance contains N atoms, the nth atom having coordinates (x_n, y_n, z_n) given as fractional parts of the lattice parameters a, b and c. The position of the nth atom in the unit cell is given by the vector r_n, where

$$\mathbf{r_n} = h\mathbf{x_n} + k\mathbf{y_n} + l\mathbf{z_n} \qquad (4.8)$$

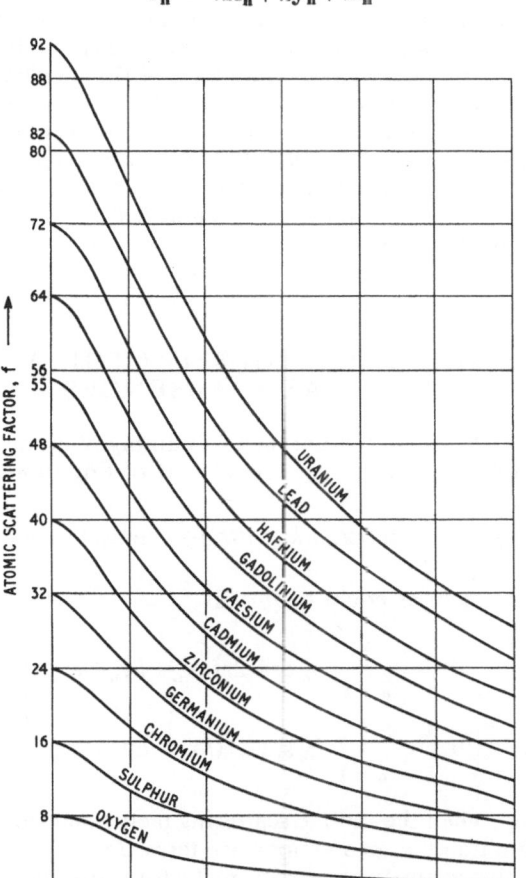

Fig. 4.4. Atomic scattering factors of various elements

75

The structure factor F is a resultant of the number of waves scattered by the atoms of the unit cell referred to the origin. The resultant F can be found by the addition of the vectorial quantities, as follows:

$$F = \sum_{n=1}^{N} f_n \exp 2\pi i (hx_n + ky_n + lz_n) \tag{4.9}$$

where f_n is the atomic scattering factor of the nth atom. The complex form of the expression indicates that the phase of the scattered waves is not simply related to the phase of the incident radiation.

The observable intensity of diffracted X-rays is found to be directly proportional to the square of the modulus of the structure factor. This modulus is called the *structure amplitude*, and is defined as the ratio of the amplitude of the radiation scattered in the order *hkl* by the contents of one unit cell to the amplitude of the radiation scattered by a single electron under the same conditions.

4.3 THE STRUCTURE FACTOR AND SYSTEMATIC ABSENCES

The structure factor F of a unit cell containing N atoms is given by Equation 4.9 above. The exponential part of Equation 4.9 may be written as:

$$\exp 2\pi i \, r_n = \cos 2\pi(hx_n + ky_n + lz_n) + i \sin 2\pi(hx_n + ky_n + lz_n) \tag{4.10}$$

Then, if F is taken as equal to $A' + iB'$,

$$\left.\begin{aligned} A' &= \sum_{n=1}^{N} f_n \cos 2\pi(hx_n + ky_n + lz_n) \\ B' &= \sum_{n=1}^{N} f_n \sin 2\pi(hx_n + ky_n + lz_n) \end{aligned}\right\} \tag{4.11}$$

Customarily, when the atomic scattering factor and the summation signs are omitted, the expressions are termed simply A and B and are called the geometrical structure factors. For crystals with a centre of symmetry, the sine term becomes zero, since for each atom with coordinates (x, y, z) there is a counter atom with coordinates $(\bar{x}, \bar{y}, \bar{z})$. Because of this, structure factor calculations for centro-symmetrical cells are greatly simplified, the equation for such cells being:

$$F = 2 \sum_{n=1}^{N/2} f_n \cos 2\pi(h\mathbf{x_n} + k\mathbf{y_n} + l\mathbf{z_n}) \qquad (4.12)$$

The following simple examples of structure factor calculations give some indication of the importance of the structure factor in X-ray diffraction work.

Consider first a primitive unit cell of the cubic system. This unit cell is centro-symmetrical and has only one atom at the origin. The fractional coordinates of this atom are (0, 0, 0), and the structure of the cell is therefore given by the equation:

$$F = f \exp 2\pi i(h0 + k0 + l0)$$

The structure factor for a primitive cubic cell is thus independent of the indices hkl and depends only on the scattering power of the atom in that particular direction.

Next consider a body centred cubic lattice of an element such as tungsten. The equivalent point coordinates are (0, 0, 0) and $(\frac{1}{2}, \frac{1}{2}, \frac{1}{2})$. Substitution of these coordinates into Equation 4.9 gives the relationship:

$$F = f[\exp 2\pi i0 + \exp 2\pi i(\tfrac{1}{2}h + \tfrac{1}{2}k + \tfrac{1}{2}l)]$$
$$= f[1 + \exp \pi i(h + k + l)]$$

But

$$\exp 2n\pi i = +1$$

and

$$\exp (2n+1)\pi i = -1$$

where n is an integer. Therefore, for $h + k + l$ even

$$F = f(1 + 1) = 2f$$

and for $h + k + l$ odd

$$F = f(1 - 1) = 0$$

From these equations, it can be seen that the only atomic planes which reflect X-rays are those whose indices add up to an even number; on all other planes, destructive interference of diffracted waves occurs, as shown in Fig. 4.5. Hence the reflections of these

77

'odd' planes will generally be absent from the spectrum; this is called the *systematic absences* phenomenon.

A careful inspection of the indices of reflections present on a photograph provides information on the space lattice symmetries of the substance under investigation. For example, if the indices 100, 111, 210, 211, etc., are systematically missing from a reflection pattern, the lattice involved is very probably body centred. In the following example, a face centred cubic lattice of an element such as copper or aluminium is considered. The equivalent point coordinates are $(0, 0, 0)$, $(\frac{1}{2}, \frac{1}{2}, 0)$, $(\frac{1}{2}, 0, \frac{1}{2})$ and $(0, \frac{1}{2}, \frac{1}{2})$. The structure factor is therefore expressed as:

$$F = f[1 + \exp 2\pi i(\tfrac{1}{2}h + \tfrac{1}{2}k) + \exp 2\pi i(\tfrac{1}{2}h + \tfrac{1}{2}l) + \exp 2\pi i(\tfrac{1}{2}k + \tfrac{1}{2}l)]$$
$$= f[1 + \exp \pi i(h + k) + \exp \pi i(h + l) + \exp \pi i(k + l)]$$

If h, k and l are either all odd or all even (unmixed), then the three sums $h + k$, $h + l$ and $k + l$ are all even and each exponential term

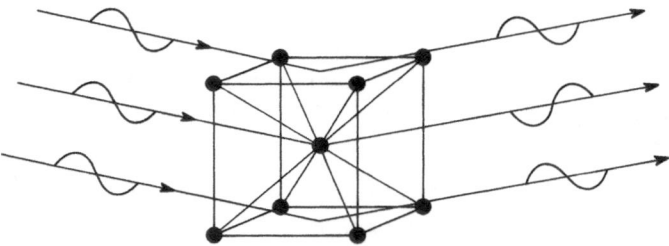

Fig. 4.5. Reflection of X-rays in the (100) plane by a body centred cubic lattice

has the value 1. (Note that a zero index is always considered as an even number.) Thus for h, k and l all odd or all even

$$F = 4f$$

and for h, k and l mixed

$$F = 0$$

A unit cell can also be base centred; this means that only one pair of opposite faces is centred (monoclinic C-face centred). If there are two atoms of the same kind per unit cell, the point coordinates are $(0, 0, 0)$ and $(\frac{1}{2}, \frac{1}{2}, 0)$. The structure factor may then be written as:

$$F = f[1 + \exp \pi i(h + k)]$$

78

As with the face centred lattice, if h and k are either both odd or both even, the exponential part of the expression has the value 1. When mixed indices are present, the exponential part becomes equal to -1. Therefore, for h and k both odd or both even

$$F = 2f$$

and for h and k mixed

$$F = 0$$

The above few simple examples and results are usually sufficient to identify Bravais lattice types from the systematic absences in powder photographs. More complicated structure factors and amplitudes may be calculated from the original structure factor equation, and can be critically examined for systematic absences.

5
The Laue method

5.1 EXPERIMENTAL TECHNIQUE

The basic equipment for the Laue method, shown diagrammatically in Fig. 5.1, is extremely simple. A beam of X-rays, about 0·5 mm in diameter, passes through a collimating system and is then allowed to fall upon a crystal. After diffraction by the crystal, the X-rays fall on a photographic plate, which records the positions and intensities of the diffracted rays.

For constructive interference of the diffracted X-rays to occur, the stringent condition of the Bragg equation $n\lambda = 2d \sin \theta$ must

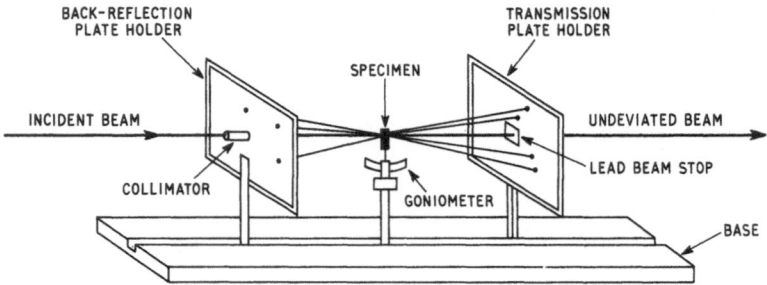

Fig. 5.1. Basic equipment for the Laue technique

be satisfied. Since the geometrical relationship between the incident beam and the diffracting atomic planes is constant, only a few planes will occupy positions which satisfy the Bragg condition. If the incident beam is composed of heterogeneous X-rays, some

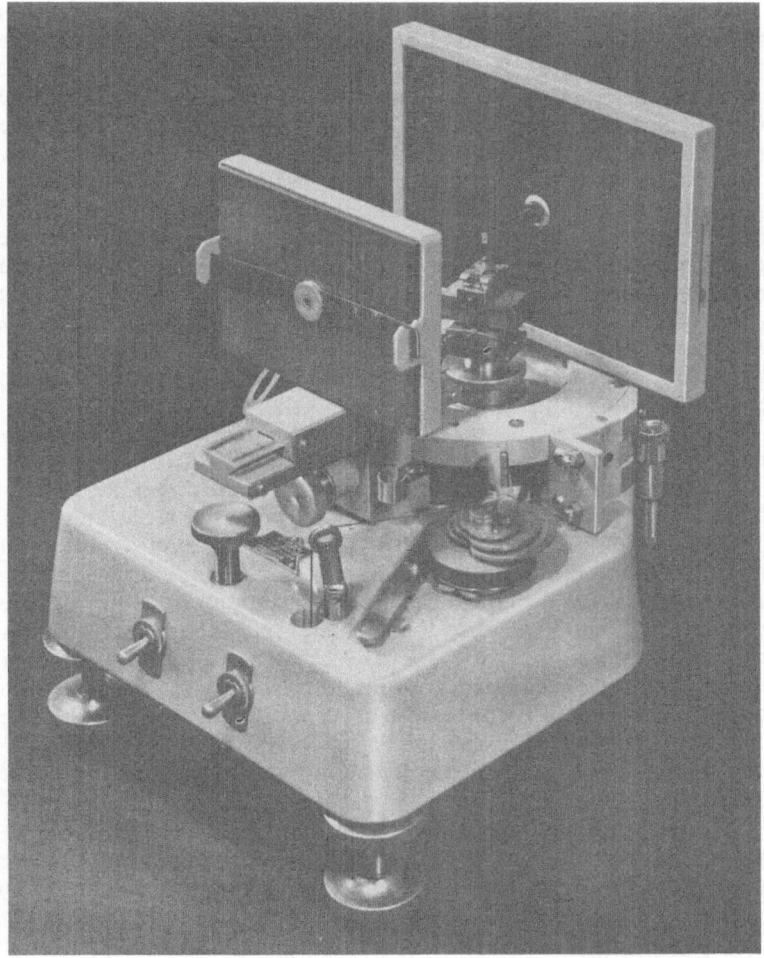

Fig. 5.2. A single-crystal X-ray camera with transmission and back-reflection plate attachments. (Courtesy Unicam Ltd)

discrete values of the wavelength λ will satisfy the Bragg equation for any orientation of the lattice planes. It appears that each particular set of atomic planes selects a suitable wavelength from the heterogeneous source and diffracts it in accordance with the Bragg law.

81

Two variations of the technique, depending on the position of the photographic plate, can be used for most purposes. The photographic plate is either placed behind the crystal for the transmission Laue method, or it is placed between the source and the specimen for the back-reflection Laue method. In routine metallurgical work when the specimen has either sufficiently low absorption or is suitably thin, transmission and back-reflection photographs are usually taken simultaneously. However, it should be remembered that good clear transmission patterns can be obtained with much shorter exposure times than back-reflection patterns. This is because of the dependence of the atomic scattering power f on the quantity $(\sin \theta)/\lambda$. As the quantity $(\sin \theta)/\lambda$ increases, the scattering factor decreases in accordance with Equation 4.7; hence the forward-scattering power of atoms is much larger than the back-scattering power.

Fig. 5.2 shows a commercial X-ray camera with the back-reflection and transmission film holders attached.

Exposure times may further be reduced by means of an intensifying screen (normally placed in contact with the recording film). This is a fluorescent screen with the ability to emit visible light under the influence of impinging X-rays. Most modern intensifying screens are based on either calcium tungstate or on zinc sulphide containing silver in traces: the former kind is most suitable for shortwave radiation, while the latter kind is designed for longer wavelengths. Intensifying screens should only be used in grain orientation studies: for other purposes, particularly when the fine details of the spots themselves are important (such as in precipitation or distortion studies), the use of intensifying screens should be avoided.

A brief examination of transmission and back-reflection photographs reveals that the diffraction spots are arranged in a definite pattern. The tautozonal planes (planes belonging to the same zone) reflect X-rays in such a way that the spots lie on ellipses or on hyperbolae in the transmission photographs, and on hyperbolae only in back-reflection photographs. This geometrical orderliness arises from the fact that, when the incident beam makes an angle θ with the zone axis, it makes the same glancing angle θ with all the faces beloning to that zone. As a result, the reflected rays will lie on the surface of a cone. The back-reflection plate then intersects the cone in a hyperbola on which the reflected spots are located. The formation of zonal lines in transmission and back-reflection photographs is illustrated in Fig. 5.3.

82

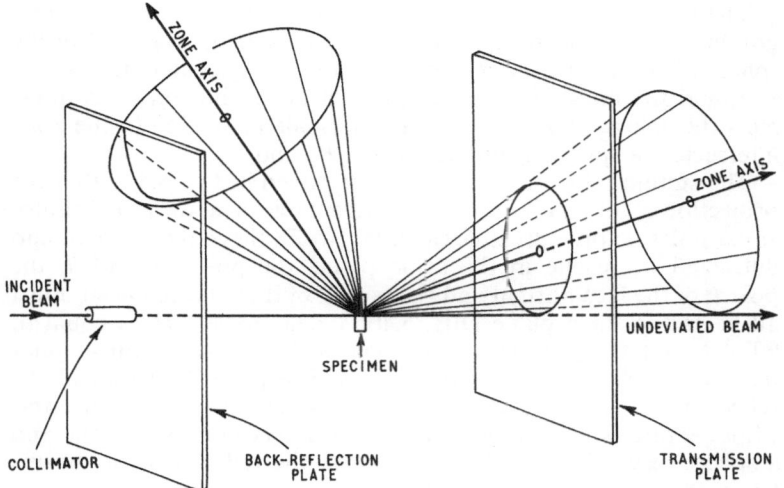

Fig. 5.3. The derivation of zonal lines in the transmission and back-reflection Laue photographs

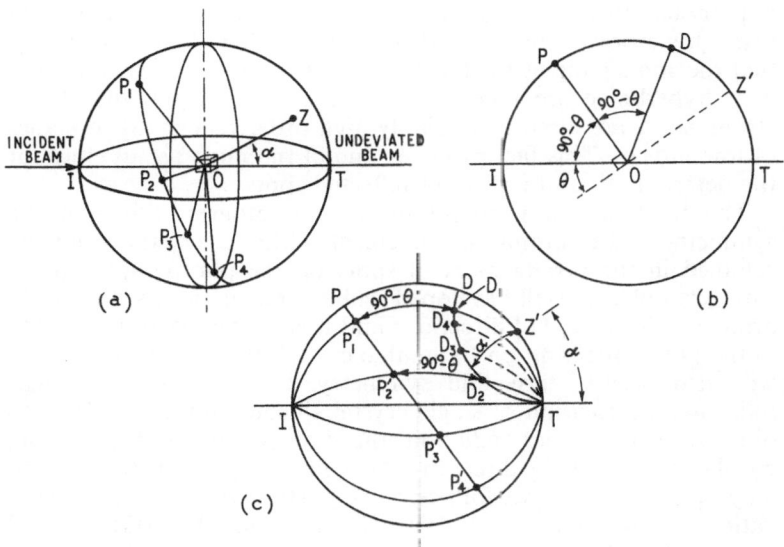

Fig. 5.4. Stereographic projection of the transmission Laue method

The Laue patterns may be demonstrated with the aid of a stereographic projection. In Fig. 5.4(a), a crystal is at the centre O of the sphere of reflection; the incident beam of X-rays enters the sphere at I, and the undeviated beam leaves it at T. The zone axis intersects the sphere at Z. A set of face normals belonging to the zone intersects the sphere at points P_1, P_2, P_3 and P_4.

The sterographic projection is so arranged in Fig. 5.4(b) that the projection Z' of Z lies on the primitive circle. The projected tautozonal poles lie on a great circle at 90° to Z'. Since the incident and diffracted beams lie in the same plane, the points I and T, the pole P of the face normal and the pole D of the reflected beam must all lie in the same plane [Fig. 5.4(b)]. The angle IOP is equal to $90° - \theta$, and the pole D of the reflected beam lies at an angular distance of $90° - \theta$ from P towards T. The poles D_1 to D_4 of the reflected beams constructed in this manner lie on a small circle whose centre is at Z' [Fig. 5.4(c)]. The intersection of a cone and a sphere results in a small circle; hence, the reflected X-ray beams lie on the surface of a cone whose axis is coincident with the zone axis.

Since crystals are composed of a number of zones, the diffraction pattern is made up of spots lying on several intersecting ellipses or hyperbolae. Inspection of back-reflection photographs shows that some hyperbolae are more densely populated than others, and that the spots lying at the intersection of two or more such prominent hyperbolae are much darker than their neighbours. These strong spots are almost invariably high-order reflections of planes of low indices. This fact provides one of the most useful clues for the determination of indices of reflected spots.

The fundamental property of Laue photographs is that the symmetry of the atomic arrangement within a crystal is largely retained in the arrangement of spots on the photographic plate. For example, a well-developed cubic crystal shows three-fold symmetry in the [111] direction; and if an X-ray beam is directed in the [111] direction on to a cubic crystal, the diffraction pattern will also possess three-fold symmetry. Fig. 5.5 shows a back-reflection photograph of single crystal silicon (cubic, $a = 5\cdot428\text{Å}$) obtained with copper radiation, the direction of the beam being parallel to the [111] direction. At a first inspection, the pattern may appear to have six-fold symmetry. However, a closer investigation reveals that the symmetry is three-fold. The three circled spots on the photograph are present only in one set of three zones, but absent from the others.

This property of symmetry retention makes the Laue method extremely useful in problems concerning grain orientations. However, the Laue method is not very suitable for the determination of atomic arrangements within the unit cell or for finding

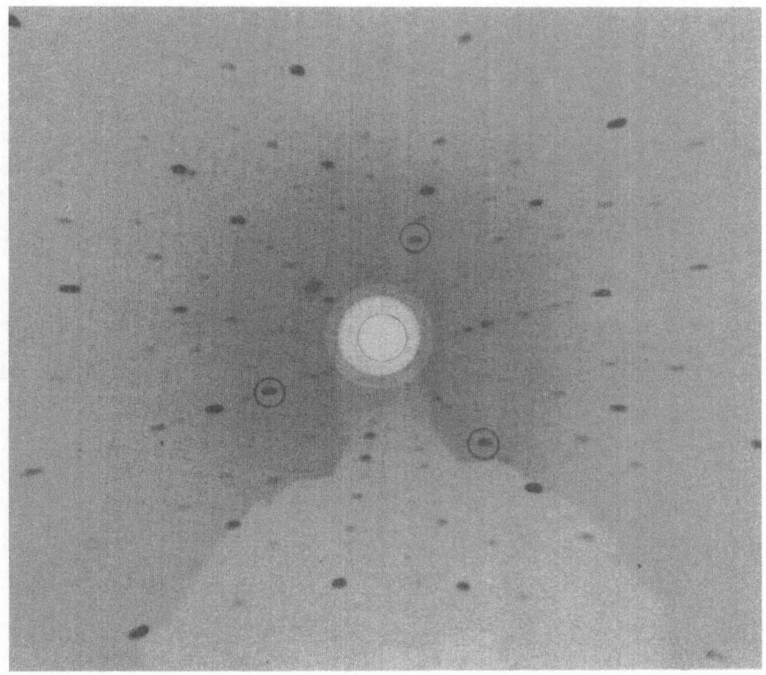

Fig. 5.5. Back-reflection Laue photograph of single crystal silicon

unknown lattice parameters — mainly because of the difficulty involved in identifying the wavelength corresponding to a specific spot. Methods more suited to such investigations will be discussed in later chapters.

5.2 INTERPRETATION OF
BACK-REFLECTION LAUE PHOTOGRAPHS

Many properties of polycrystalline materials have been studied by examinations of single isolated grains. Mechanical, electrical

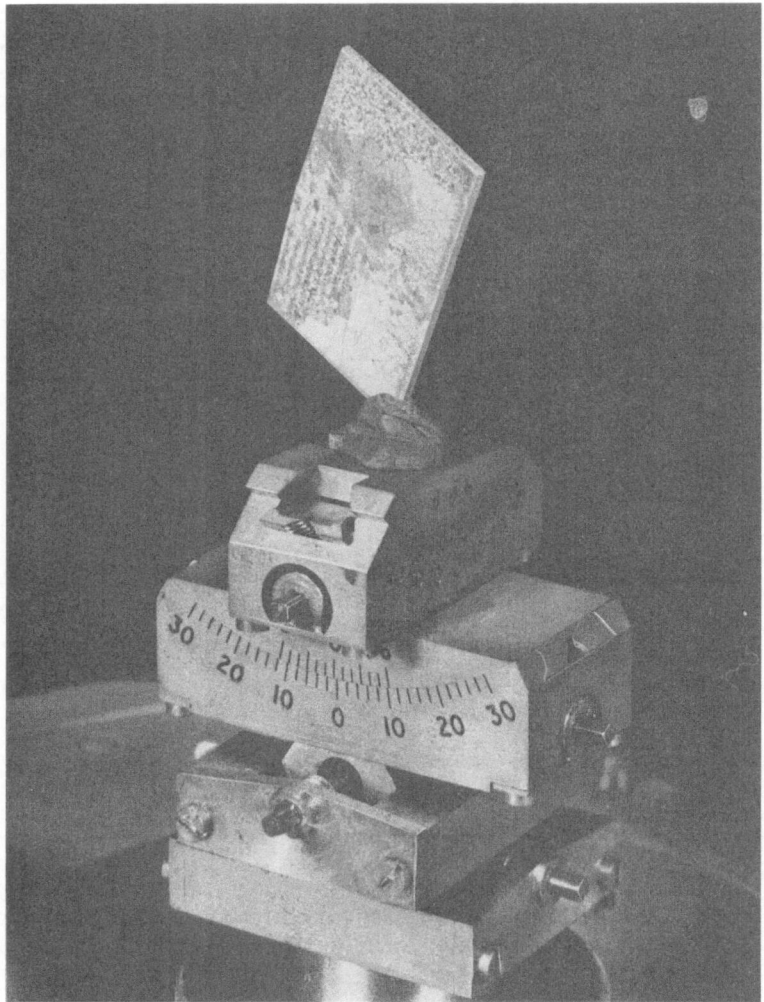

Fig. 5.6. A goniometer head. (Courtesy Unicam Ltd)

or chemical behaviour of metals depends on the specific directions in which these properties are measured. The aim should be to obtain an accurate determination of at least one crystallographic axis of a grain, relative to some reference plane or direction. With

86

Table 5.1.
Angles in Cubic Crystals between Planes of the Form $\{h_1k_1l_1\}$ and $\{h_2k_2l_2\}$

$\{h_2k_2l_2\}$	$\{h_1k_1l_1\}$						
	(100)	(110)	(111)	(210)	(211)	(221)	(310)
(100)	0° 90°						
(110)	45° 90°	0° 60° 90°					
(111)	54·74°	35·3° 90°	0° 70·5°				
(210)	26·6° 63·4° 90°	18·4° 50·8° 71·6°	39·2° 75·0°	0° 36·9° 53·1°			
(211)	35·3° 65·9°	30° 54·7° 73·2° 90°	19·5° 61·9° 90°	24·1° 43·1° 56·8° 79·5°	0° 33·6° 48·2° 60·0°		
(221)	48·2° 70·5°	19·5° 45° 76·4° 90°	15·8° 54·7° 78·9°	26·6° 41·8° 53·4° 63·4°	17·7° 35·3° 47·1° 65·9°	0° 27·3° 38·9° 63·6°	
(310)	18·4° 71·6° 90°	26·6° 47·9° 63·4° 77·1°	43·1° 68·6°	8·1° 31·9° 45° 54·9°	25·4° 40·21° 58·9° 75·0°	32·5° 42·5° 58·2° 65·1°	0° 25·9° 36·9° 53·1°
(311)	24·2° 72·5°	31·5° 64·8° 90°	29·5° 58·5° 80·0°	19·3° 47·6° 66·1°	10·0° 42·4° 60·5°	25·2° 45·3° 59·8°	17·6° 40·3° 55·1°
(320)	33·7° 56·3° 90°	11·3° 54·0° 66·9° 78·7°	36·8° 80·8°	7·1° 29·8° 41·9°	25·1° 37·6° 55·5°	22·4° 42·3° 49·7°	15·3° 37·9° 52·1°
(321)	36·7° 57·7° 74·5°	19·1° 40·9° 55·5° 67·8°	22·2° 51·9° 72·0° 90°	17·0° 33·2° 53·3° 90°	10·9° 29·2° 40·2° 90°	11·5° 27·0° 36·7° 57·7°	21·6° 32·3° 40·5° 47·5°
(331)	46·5°	13·3°	22·0°	22·6°	20·5°	6·21°	29·5°
(510)	11·3°	33·7°	47·2°				
(511)	15·8°	35·3°	38·9°				

a flat specimen, the reference plane may be chosen as the surface of the grain; with a cylindrical specimen, the longitudinal axis is suitable.

The sample is normally set up in a goniometer head (Fig. 5.6), which is provided with some mechanical means of rotation about three perpendicular axes and translation along the same axial

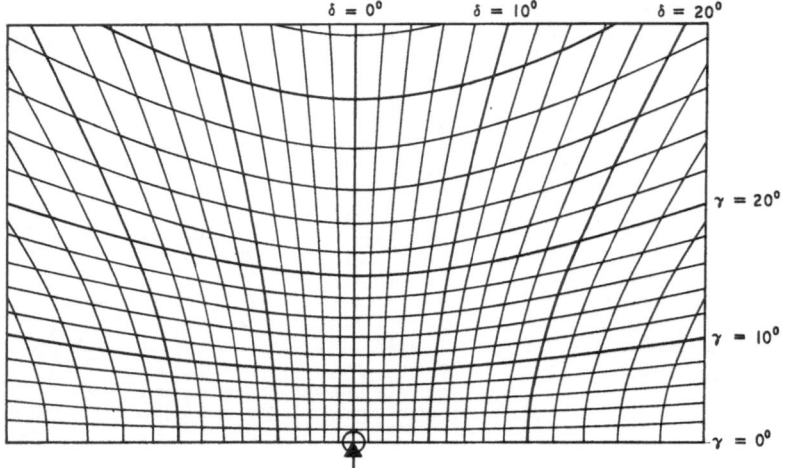

Fig. 5.7. *The Greninger chart for a specimen-to-film distance of 3 cm*

directions. The goniometer should be precision made, so that the specimen can, if necessary, be moved and then returned to its exact original position. Reasonable equipment will enable an experienced operator to determine the orientation of a specimen with an accuracy between $0.5°$ and $1.0°$.

The Laue camera is constructed in such a way that the positions of the photographic plate holders bear some simple geometrical relationship to the direction of the X-ray beam: normally they are placed perpendicular to the incident X-ray beam. The specimen is then placed parallel to the photographic plate. The relationship between the Laue spots and the incident X-ray beam must then be found in order to fix the orientation of some specific crystallographic direction.

It was shown in Section 5.1 that tautozonal atomic planes reflect an incident beam to give diffraction spots whose locus is a conic section. If it can be shown that some spots belong to tautozonal

88

planes, the angle between the zone axes may be measured. Once a set of angles has been measured, their indices may be identified by comparison with a table of known interplanar angles for the crystal system involved. Table 5.1 lists selected values of interplanar angles for the cubic system.

A back-reflection Laue pattern can easily be interpreted with the aid of the Greninger chart (Fig. 5.7). This chart comprises two sets of hyperbolae: one set measures the acute angle between the incident and diffracted beams, and the other reads the angles between crystal planes. The Greninger chart may be used in a number of ways, but only two basic methods will be described here.

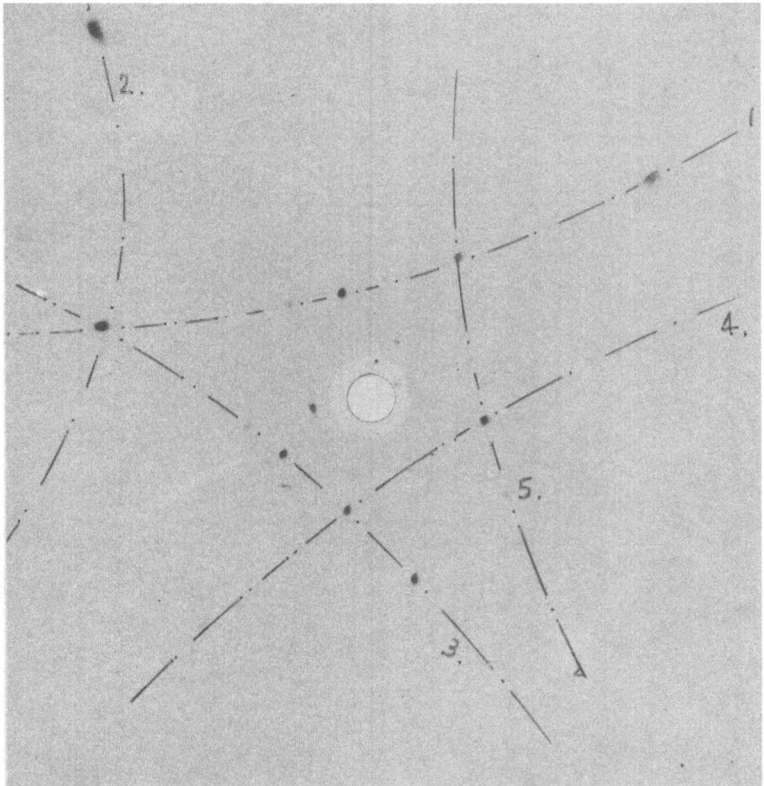

Fig. 5.8. Back-reflection Laue photograph of single crystal aluminium, with some zone hyperbolae marked

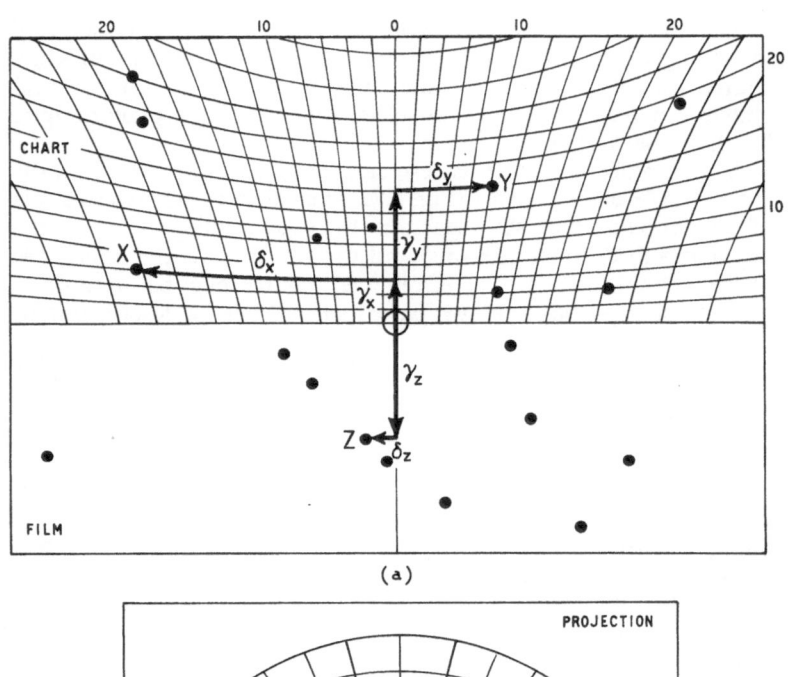

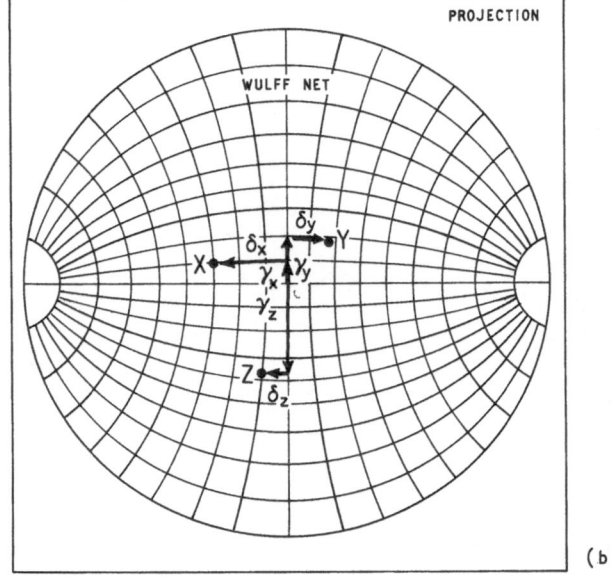

Fig. 5.9. *Application of the Greninger chart to plotting reflections of planes*

Fig. 5.8 shows a back-reflection Laue photograph of single crystal aluminium (cubic, $a = 4\cdot0490$ Å) obtained with copper radiation at 35 kV and 13 mA and an exposure time of 45 min. In Fig. 5.9(a), the Greninger chart is superimposed on the photograph of Fig. 5.8. Three obviously important spots have been chosen and labelled X, Y and Z. (These spots were selected on the grounds that their intensities are higher than those of neighbouring spots, and that they lie at the intersections of several zone hyperbolae.) A set of coordinates γ_x, γ_y and γ_z, representing the angles between the crystal planes and an imaginary reference position, are first measured. Then another set of coordinates are measured: δ_x, δ_y and δ_z, representing the acute angles between the incident beam and each diffracted beam. The three selected points are next plotted on a stereographic projection, as shown in Fig. 5.9(b). In order to measure the angles between consecutive face normals, the Wulff net is rotated so that any two poles lie on the same great circle.

The angles measured in this instance are:

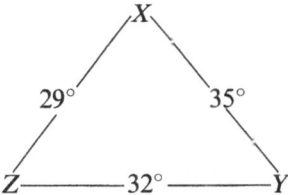

From Table 5.1, it can be seen that the angle $\widehat{XY}$ corresponds to one possible angle between the (110) and (111) face normals. However, the table does not show which index belongs to either of the poles — it only gives a possible pair.

Assume that X is the pole (110) and Y is the pole (111). In order to determine the indices of pole Z, the plane which lies at 29° to the (110) plane and at 32° to the (111) plane must be found. However, it is soon evident from Table 5.1 that there is no plane lying at these angles. The order of poles X and Y must therefore be reversed — the 111 indices being assigned to pole X, and the 110 indices to pole Y. From Table 5.1, it can be seen that the (311) plane lies at an angle of 31·5° to the (110) plane and 29·5° to the (111) plane. Since these values closely correspond to the measured angles, the indices of the poles X, Y and Z can be taken as 111, 110 and 311 respectively.

91

Now that the indices of the reflections are known, the orientation of the specimen can be given as an angular dimension between one crystallographic direction, say [111], and the direction of the incident beam [001]; in this instance, the orientation is 25°. It is sometimes convenient to give the indices of the incident beam in terms of the crystal, stating that the incident beam is parallel to the [uvw] direction of the crystal. In this example, the incident beam is perpendicular to the plane of the stereogram and is almost parallel to the [321] direction of the crystal.

The success of this method of interpreting back-reflection photographs depends mainly on the condition of the sample and of the diffracted spots. The spots must be clear and their positions easy to measure; there must also be a clear indication that they are diffracted by low-index planes. Because of thermal vibrations or stress distortions, the reflections are often blurred and their positions are not clear. The recommended alternative treatment of back-reflection photographs is to identify the zone axes of prominent tautozonal spots and to plot them on a stereogram. This procedure is used here to locate the orientation of a single crystal of silicon (cubic, $a = 5\cdot4282$ Å), whose Laue photograph is shown in Fig. 5.10(a). The photograph was taken with unfiltered copper radiation and an exposure time of 45 min, and some selected spots have been traced in Fig. 5.10(b).

First, some prominent lines are selected and numbered for identification purposes [Fig. 5.10(b)]. Fig. 5.11 shows the principle of plotting one tautozonal hyperbola and its zone axis in stereographic projection, and Fig. 5.12 illustrates the complete set of hyperbolae and the poles of their zone axes in stereographic projection. The next step is to measure the angles between the zone axes and to compare them with calculated angles between crystallographic planes.

The observed and calculated values are as follows:

Poles	Observed angles	Zone axes	Calculated angles
$\widehat{P_1P_2}$	59·3°	[322][331]	59·95°
$\widehat{P_2P_3}$	76·0°	[331][100]	76·74°
$\widehat{P_3P_4}$	48·8°	[100][221]	48·19°
$\widehat{P_1P_3}$	60·2°	[322][100]	60·98°

Examination of several combinations shows that pole P_3 is the

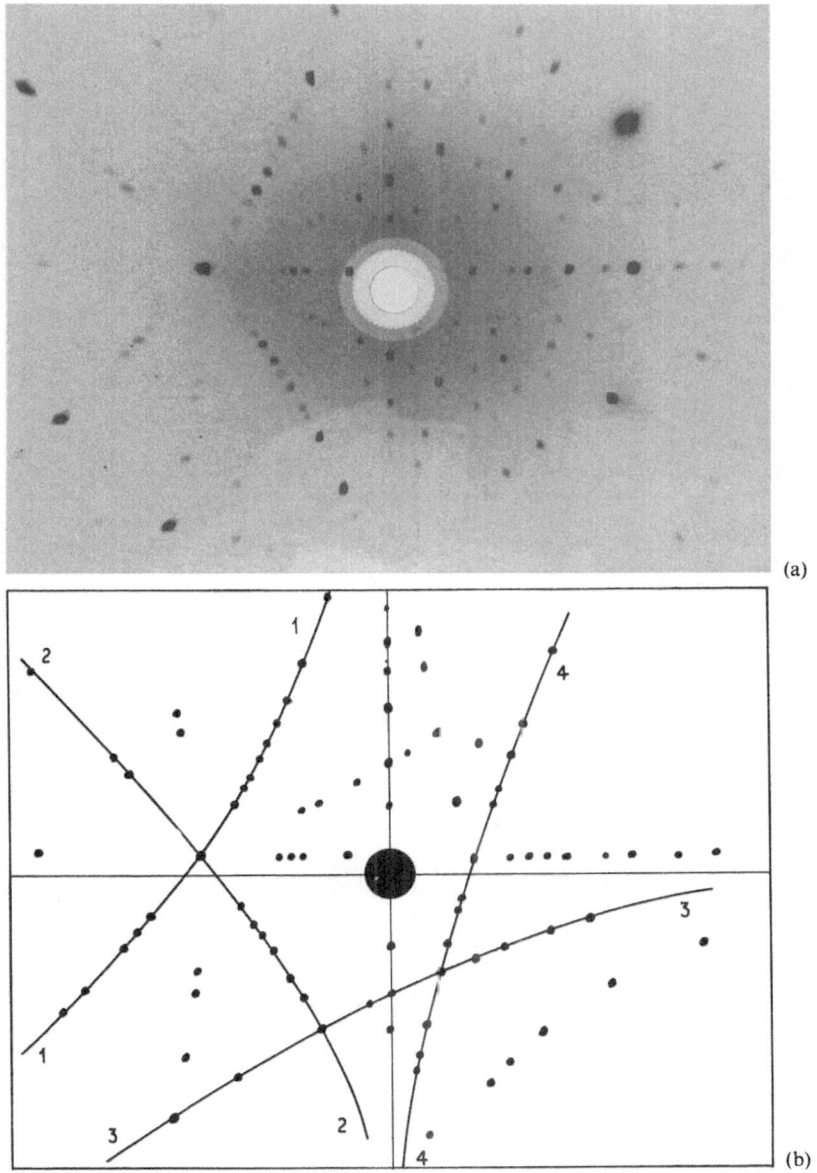

Fig. 5.10. (a) Back-reflection Laue photograph of single crystal silicon; (b) trace of selected spots, with some zone hyperbolae marked

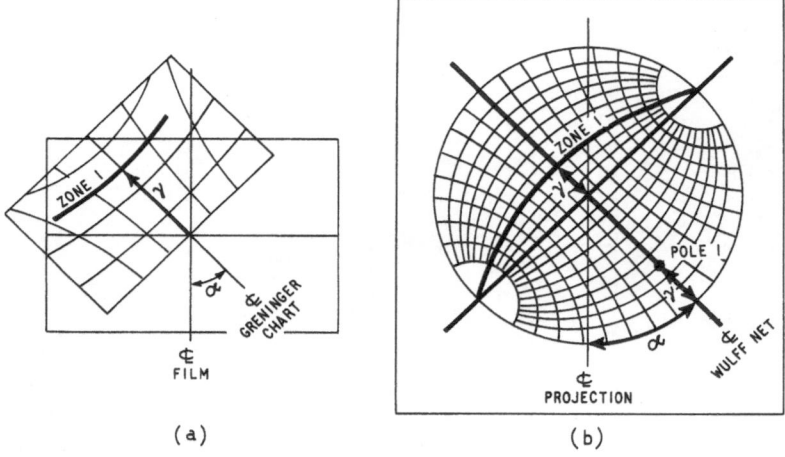

(a)   (b)

Fig. 5.11. Application of the Greninger chart to plotting a zone and the pole of its zone axis on a stereographic projection

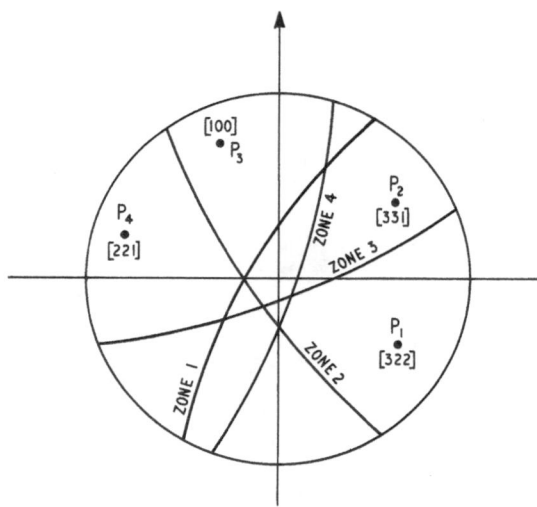

Fig. 5.12. The stereographic projection of zones indicated in Fig. 5.10(b)

[100] zone axis, and hence all the other poles may be indexed. Thus the orientation of the crystal is defined.

For some types of investigation, a diffraction pattern obtained with the crystal in a specified position relative to the incident beam is needed. The three-circle goniometer on which the specimen is mounted can be used to bring the crystal into the required position. For example, assume that the [100] zone axis of the silicon crystal used above must be parallel to the incident X-ray beam, and that its [110] zone axis should point to the right (looking in the direction of the incident beam). The required orientation may be achieved by three rotations, as shown in Fig. 5.13

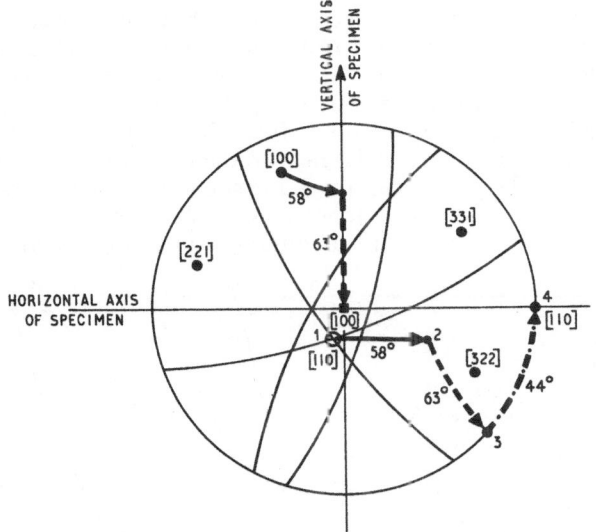

Fig. 5.13. Stereographic plot of the rotations needed to bring the specimen into the required orientation

(which is a stereographic projection of the pattern in Fig. 5.10, with the pole of the [110] zone indicated). First, a small circle rotation of 58° brings the pole [100] on to the vertical great circle. At the same time, the pole [110] also moves on a small circle from position 1 to position 2. The second rotation of 63° on a great circle brings the pole [100] into position parallel to the incident X-ray beam, while the pole [110] moves on a small circle to position 3. The pole [110] is brought into its final position by rotation through 44° around the centre of the stereogram.

95

The stereographic rotations always correspond to actual rotations on the goniometer. The first rotation of 58° is around the vertical axis of the specimen; the second corresponds to a rotation of the goniometer around the horizontal circle perpendicular to the incident beam; and the third rotation takes place around the goniometer axis parallel to the incoming beam.

5.3 INTERPRETATION OF
TRANSMISSION LAUE PATTERNS

Transmission Laue patterns are obtained when the photographic plate is placed behind the specimen, and diffraction occurs on the side of the specimen furthest from the impinging X-rays. As with back-reflection photographs, the symmetry of atomic lattices is retained in transmission Laue patterns. Reflections originating from tautozonal planes now lie on a full ellipse in the photograph, provided that the zone axis lies near to the direction of the incident beam. In all other instances, the Laue spots lie on hyperbolae.

The interpretation of transmission photographs is greatly facilitated by the use of the Leonhard chart (Fig. 5.14). This is

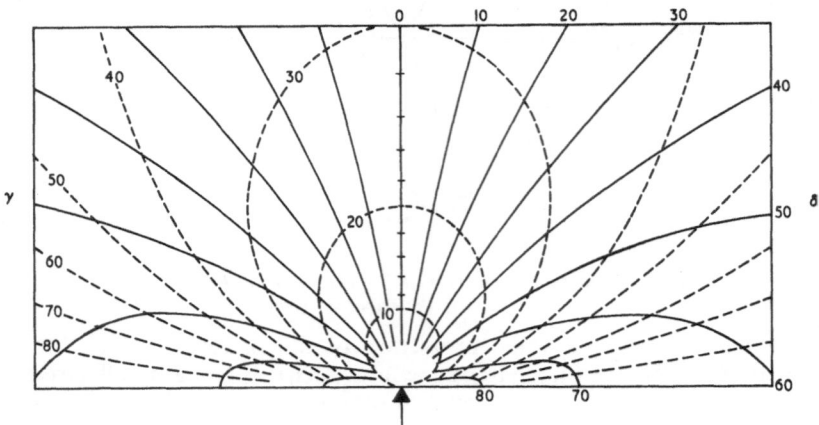

Fig. 5.14. The Leonhard chart for a specimen-to-film distance of 3 cm

similar to the Greninger chart and is used in precisely the same way. It is composed of two sets of curves: the set of dotted lines of constant γ corresponds to the elevation of great circles on the

96

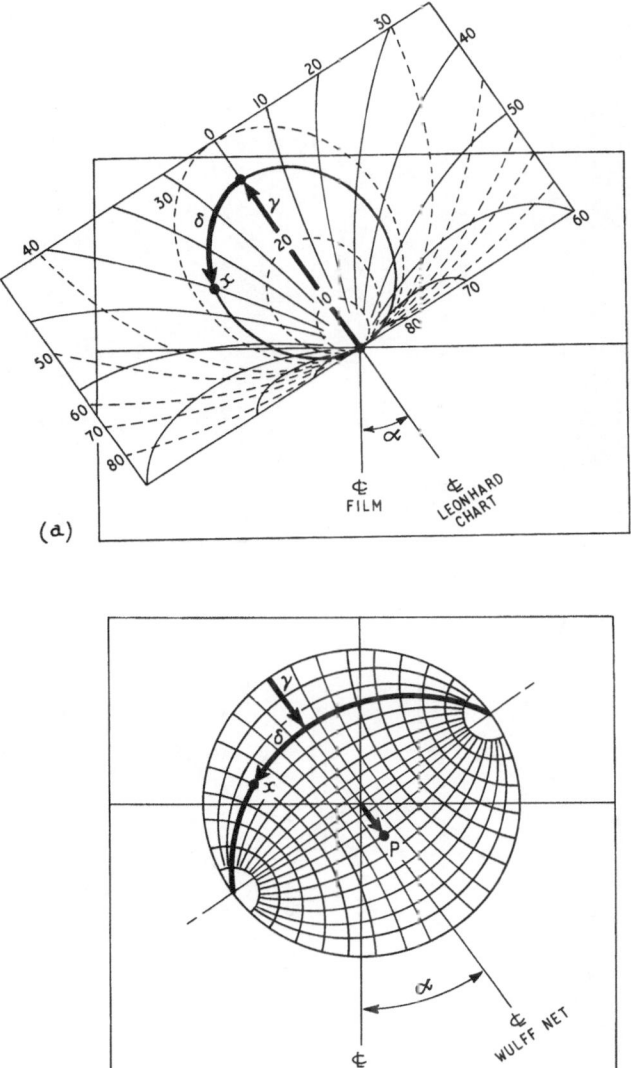

Fig. 5.15. Application of the Leonhard chart to plotting the pole of a
plane and the zone to which pole x belongs

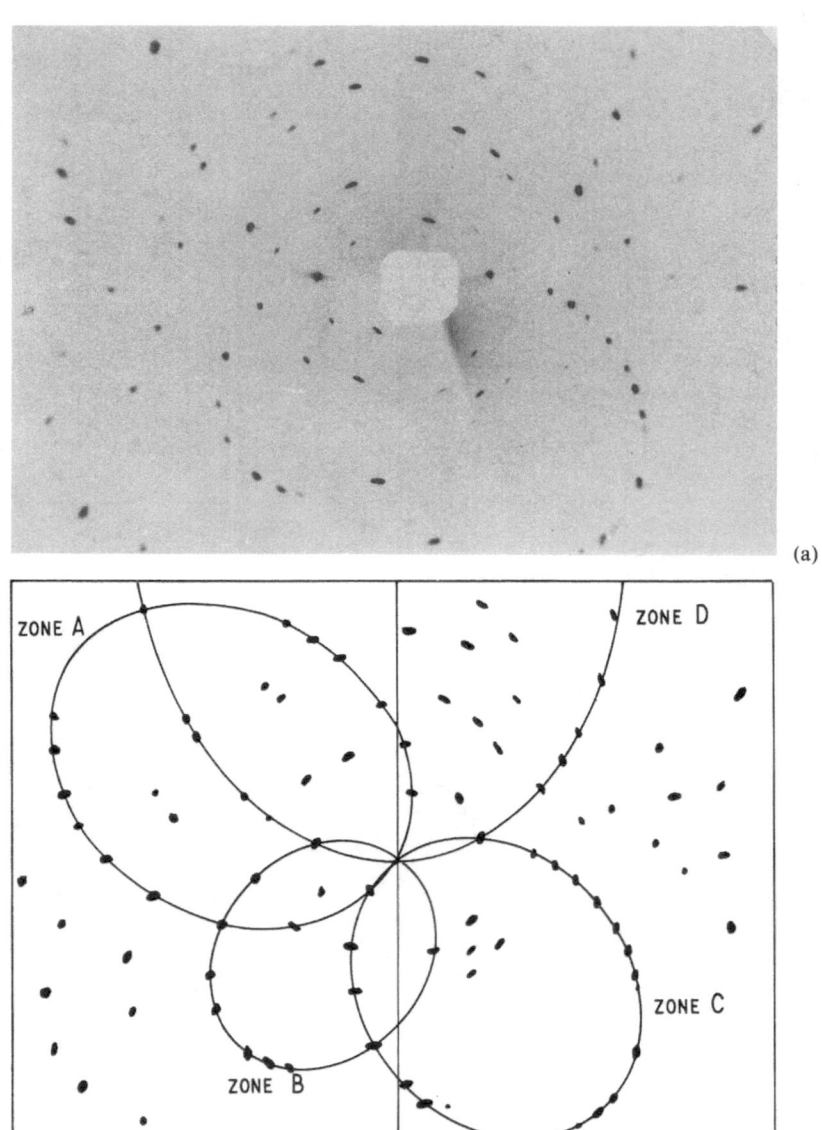

ZONE A

ZONE D

ZONE B

ZONE C

(a)

(b)

Fig. 5.16. (a) Transmission pattern of single crystal silicon;
(b) selected tautozonal spots

stereographic projection. The set of solid lines are of constant δ and correspond to the latitude lines on the Wulff net. The use of the Leonhard chart is illustrated in Fig. 5.15, but its application is again best demonstrated with an example of indexing a transmission pattern.

Fig. 5.16(a) shows a transmission pattern of single crystal silicon (cubic, $a = 5.4282$ Å) irradiated with copper radiation at 35 kV and 13 mA and an exposure time of 60 min. A number of prominent zone ellipses are clearly visible. In this example, the zones and their associated poles will be plotted, and the observed angular measurements will then be compared with calculated data. The reader may, for practice, select three or more prominent spots and use the alternative method explained in the previous section.

First, four prominent sets of tautozonal spots are selected, as indicated in Fig. 5.16(b). The necessary angular values can then be measured with the aid of the Leonhard chart and are found to be as follows:

Zone symbol	α	γ
A	57°	30°
B	137°	23°
C	213°	27°
D	2°	38°

If this set of data is plotted on the stereographic projection (Fig. 5.17), a set of measurements can be taken between the plotted poles. Thus:

$$\widehat{P_A P_B} = 33° \qquad \widehat{P_C P_D} = 60°$$
$$\widehat{P_B P_C} = 30° \qquad \widehat{P_D P_A} = 30°$$

The clue to finding the correct indices is in the 60° angle between P_C and P_D. For a first trial, assume that both P_C and P_D are [110] directions. The pole P_B, which lies at 30° to P_C, may then be the [211] direction. If this direction is correct, the pole P_C must lie at one of the following angles to the pole P_D: 30·0°, 54·74°, 73·22° or 90·0°. The actual measurement between P_B and P_D is 54°, which confirms the assumption (within the error of the graphical construction) that P_B is the [211] direction. If this argument is continued, the indices of all the other directions may be found.

The direction of the incident X-ray beam in terms of the crystal indices may now be found. As poles P_C and P_D are both [110]

99

directions, the centre of the stereogram (the pole of the incident beam) lies at 23° and 38° to poles P_C and P_D. Consulting the tables of angles, the pole of the [432] direction lies at 23·20° and 38·02°

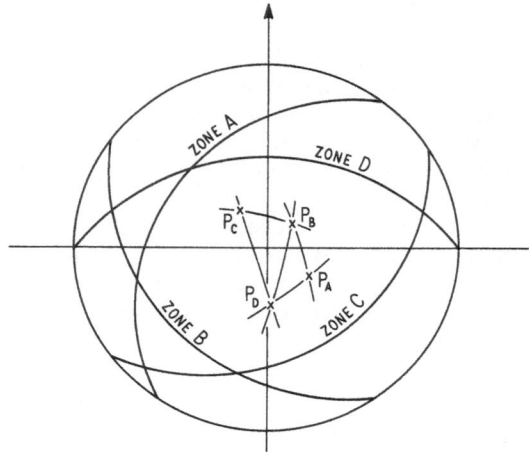

Fig. 5.17. Stereographic projection of the zones shown in Fig. 5.16(b)

to the direction [110]. This now completes all the information necessary to define numerically the orientation of the silicon specimen.

5.4 INTERPRETATION OF THE SHAPES OF LAUE SPOTS

Both transmission and back-reflection Laue photographs contain valuable information regarding the state and nature of the crystal under investigation. It can be seen from Fig. 5.16(a), for example, that the actual transmission spots are not circular but rather elliptical in shape. The ratio of the major to minor axes of each ellipse varies with the position of the spot. The nearer the spot is to the centre of the photograph, the more circular it appears to be. Further from the centre, the major axes of the elliptical spots align themselves in a specific direction, often in line with the tautozonal ellipse.

The effect of divergence of the incident X-ray beam on a theoretically perfect strain-free crystal will first be considered. It is

important to realise this effect of divergence, or otherwise a somewhat irregular shape of spot may be erroneously attributed to
crystal distortion. Fig. 5.18 shows a section through an incident

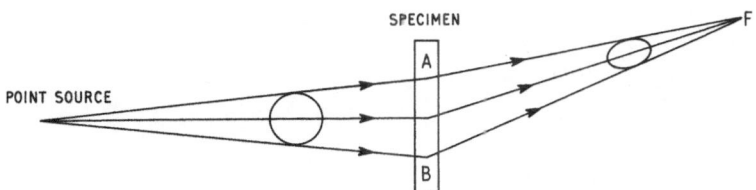

Fig. 5.18. *Effect of divergence on the shape of a Laue spot*

and diffracted X-ray beam. Each ray of the circularly divergent
incident beam strikes the reflecting lattice planes at a slightly
different angle. Since the ray striking the crystal at A is reflected
through a greater angle then the ray striking at B, an effect of
focusing will occur at point F. Those rays travelling in the plane
perpendicular to the plane of the drawing will continue to diverge,
after diffraction, without angular variation. The result is an elliptical cross-section of the diffracted beam. As the film intersects the
various diffracted beams at different angles and distances from the
film, the elliptical spots appear to be of various sizes, as observed
in Fig. 5.16(a).

The effect of stress on Laue spots is best understood by investigating the changes it can cause in the atomic lattice pattern. So
that continuity of a crystal can be maintained, externally imposed
stresses are balanced by internal stresses, which in turn are balanced
by the interatomic cohesive forces. The applied stress may, in some
instances, change the lattice parameters slightly in one or more
directions and cause small curvatures in certain atomic planes.
Undistorted annealed metal acts like an optically flat mirror, de-
flecting the incident beam in a number of sharp spots; deformed
crystals, however, with their curved lattice planes, diffract the X-ray
beam in elongated streaks which are recorded on the photographic
plate. This effect is illustrated in Fig. 5.19 — a transmission photograph of p-toluidine taken with copper radiation.

This streaking of Laue spots is often referred to as deformation
asterism. Asterism of Laue spots is a result of changes in reflection
geometry caused by minute variations in the reflecting atomic
planes. Consider, for example, the crystal plane of Fig. 5.20 deformed so that the face normal ON lies on the surface of a cone

101

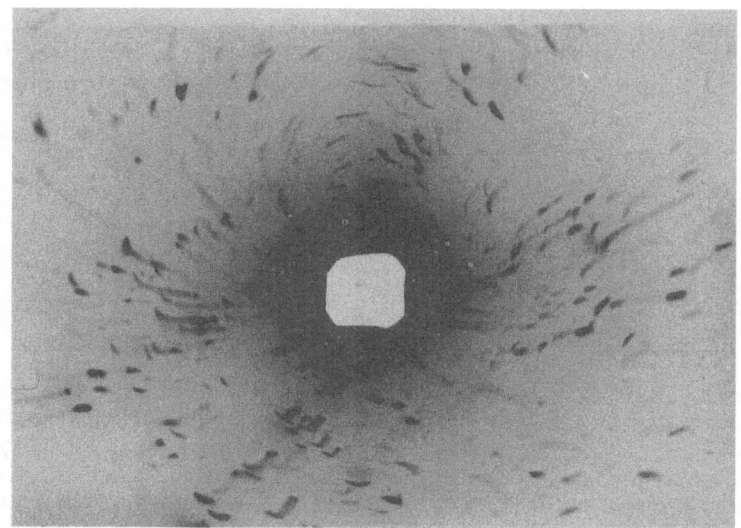

Fig. 5.19. Transmission pattern of p-toluidine

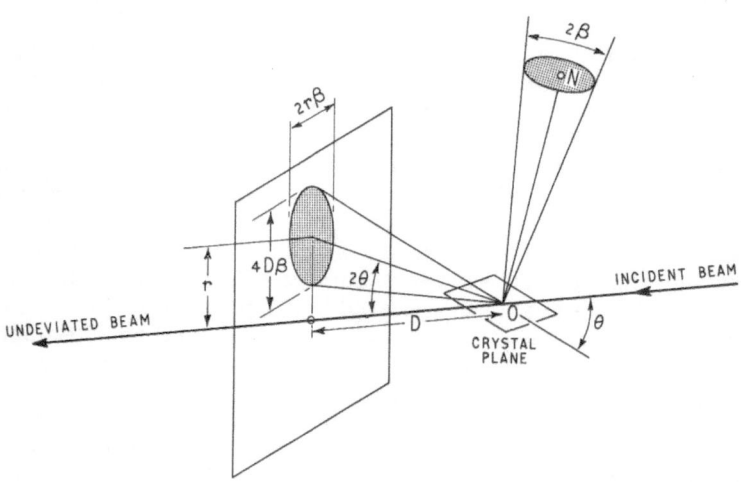

Fig. 5.20. The formation of asterism

whose apex angle is 2β radians. According to the principles of diffraction optics, the incident beam, the face normal and the reflected beam all lie in the plane of incidence. Also, if the face normal sweeps through an angle of 2β, the reflected beam will sweep through an angle of 4β in the plane of incidence. Therefore, the length of the spot in the plane of incidence will be $4D\beta$, where D is the distance between the specimen and the film. The width of the spot normal to the plane of incidence is $2r\beta$, where r is the height of the centre of the spot. If the second-order effects resulting from the fact that the film does not intersect the reflected beam perpendicular to its axis are ignored, then it follows from geometrical considerations that

$$2r\beta = 2(2D\theta)\beta = 4D\theta\beta$$

where θ is the glancing angle in radians. Therefore the ratio R of the width to the length of the spot is:

$$R = \frac{4D\theta\beta}{4D\beta} = \theta$$

The significance of this result is that the relative position of the spot in the film affects the ratio of minor to major axes very considerably. The more elliptical spots may not always represent the more severely deformed atomic lattice planes; but they do show that the mean position of the appropriate face normals lies nearer to the direction of the incident beam.

Considerable caution is needed before conclusions can be drawn about the magnitude of the deformation affecting a specific asterised spot. A spot relating to an ideally flat atomic plane is the result of the diffraction of one specific wavelength of the continuous spectrum appropriate to the spacing and the glancing angle simultaneously. Once the plane is bent, the resulting variation in the glancing angle θ means that the diffracted wavelength widens into a band of values. Occasionally the complete band of wavelength values is not available in the spectrum because part of the band lies beyond the cut-off wavelength. The result of this is that the full length of the streak is not recorded on the photographic plate.

Another strain effect can also be recorded by the Laue technique. When the specimen is part of a polycrystalline aggregate, a small stress, well within the elastic limit of the specimen, may cause a small rotation of grain. As the Laue spots are very sensitive to small positional changes, a rearrangement of spots can occur. A double exposure on the same film is necessary to record the original and the stressed specimen, and thus to show the doubling of spots.

If the deformation of plastically bent crystals is not excessive, segments of dislocations in opposing directions may align themselves so that some stress recovery occurs. The full details of this process have not yet been determined, but it seems that in the final stages the bent crystals assume a fine structure after a recovery anneal which does not recrystallise the specimen. It appears that the distorted crystal breaks up into smaller blocks, which are largely strain free. These primary blocks are disorientated by about the same amount as the bent crystal from which they originate. This process is known as polygonisation. The effect of polygonisation on Laue spots is illustrated in Fig. 5.21. The realignment of dislocations during polygonisation causes the continuous asterisms to be replaced by a set of isolated small sharp bands.

The Laue technique also provides an extremely powerful tool for solid-state precipitation studies. When a second phase is precipitated from a supersaturated solid solution, the second phase frequently forms thin plates parallel to a low-index plane of the matrix. This low-index plane of the matrix is called the *habit plane* of the second phase, and the indices are always computed in terms of the matrix grain. If the solid solution has a general (*hkl*) habit, plane precipitation of the second phase can occur on any planes of

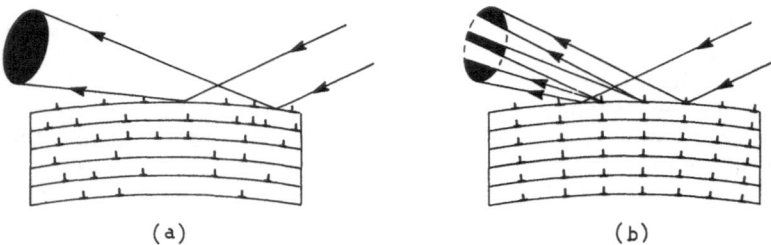

(a)   (b)

Fig. 5.21. The effect of realignment of dislocations during polygonisation on the shape of Laue spots: (a) before polygonisation; (b) after polygonisation

the form {*hkl*}. In the more practically important cases, however, precipitation is restricted to special planes, such as the (100), (110) or (111) planes. When grains of this description are prepared for microscopic examination, the thin plate-like second-phase particles appear as needles and flakes on the plane of polish — resulting in a structure customarily called the Widmanstätten structure.

104

Fig. 5.22 shows a typical Widmanstätten structure present in Al–4% Cu aged for 90 min at 350°C. The plane of section is the (100) plane. The perpendicular precipitate lines are parallel to the (010) and (001) planes, and the darker patches are plates of precipitate lying in the plane of section of the grain. The magnification

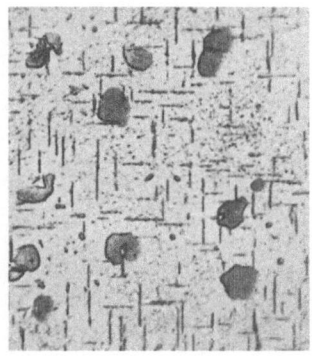

Fig. 5.22. A micrograph of Al–4% Cu aged for 90 min at 350°C. The plane of section is (100) and the magnification 2,500. [From GAYLER and PARKHOUSE, J.Inst. Metals, **66**, 67 (1940) by permission of The Institute of Metals]

of the micrograph is 2,500. This high magnification reveals the difficulty inherent in traditional microscopic investigations, which have mainly concentrated on the final stages of the reaction when the agglomeration of the second phase becomes visible under the microscope.

The incidence of age-hardening may be traced back to much earlier stages by means of X-ray investigations of single crystals of some age-hardening alloys. Fig. 5.23 shows a set of Laue photographs of the changes occurring in the process of age-hardening of Al–4% Cu alloys. Fig. 5.23(a) shows, for reference, a Laue transmission photograph of pure aluminium, the incident X-ray being parallel to the [100] direction, and the [110] direction of the crystal being vertical. A polychromatic radiation from a silver target was used for all the photographs, and the orientation and exposure data were kept constant throughout the experiment.

The crystal which gave the photograph of Fig. 5.23(b) was an Al–4% Cu crystal aged for 3 days at room temperature. The streaks and diffuse spots which appear in Fig. 5.23(b) but not in Fig. 5.23(a) are very probably due to the presence of a second phase, consisting of small plates of copper-rich material. The streaks of the central region of Fig. 5.23(b) give a convenient measure of the plates present: the plate widths are about six atomic layers (lattice parameters in that direction), and their thickness is about one or

105

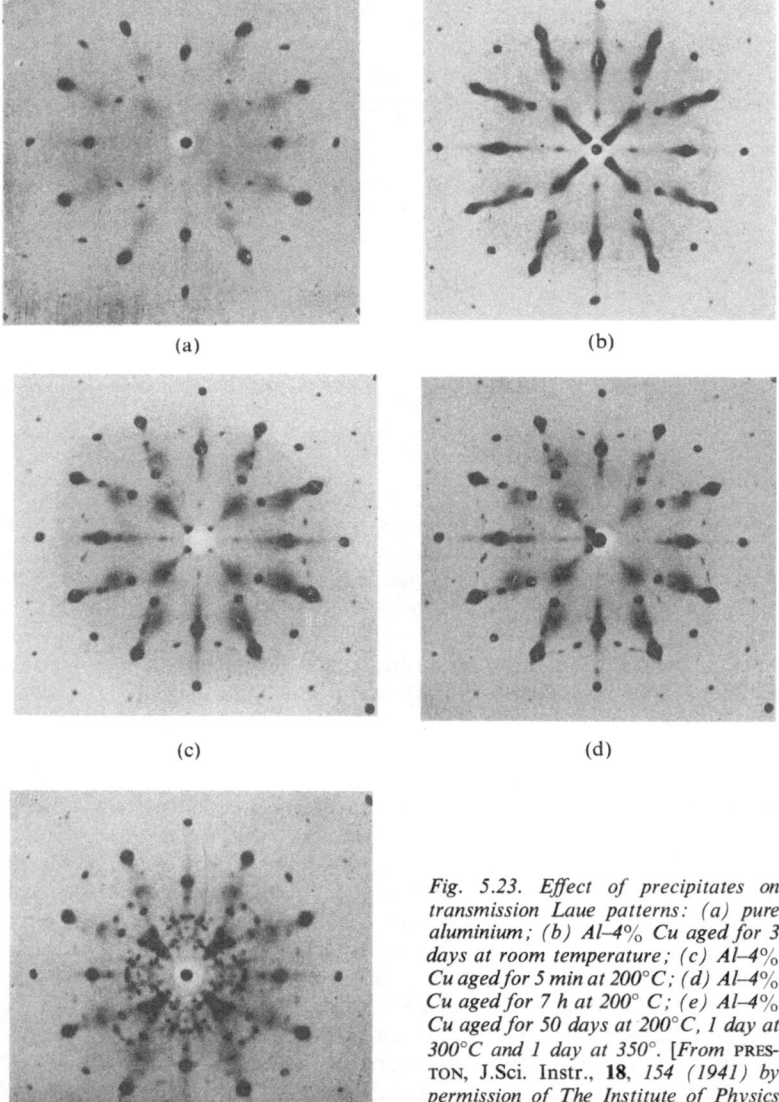

(a)

(b)

(c)

(d)

(e)

Fig. 5.23. Effect of precipitates on transmission Laue patterns: (a) pure aluminium; (b) Al–4% Cu aged for 3 days at room temperature; (c) Al–4% Cu aged for 5 min at 200°C; (d) Al–4% Cu aged for 7 h at 200° C; (e) Al–4% Cu aged for 50 days at 200°C, 1 day at 300°C and 1 day at 350°. [*From* PRES-TON, *J.*Sci. Instr., **18**, *154 (1941) by permission of The Institute of Physics and The Physical Society*]

two atomic layers. When the age-hardening is induced at 200°C, as in Fig. 5.23(c), the shortening of the central streaks shows the increase in the area of the precipitate, while the appearance of sharp outer spots shows the increase in thickness. After keeping the specimen for 7 h at 200°C, the spots are somewhat better developed — showing a slight increase in thickness of the precipitated second phase, but without appreciable change in the area as shown by the unaltered central region [Fig. 5.23(d)]. Further extensive heat treatment causes the formation of an electron compound of $CuAl_2$ whose lattice fits into the lattice of the aluminium matrix. The tetragonal symmetry of $CuAl_2$ is easily distinguishable in Fig. 5.23(e).

6
X-ray powder photography

6.1 THE PRINCIPLES OF
POWDER PHOTOGRAPHY

It is often not possible to prepare single crystals of metals or alloys and, because of the limitations of the Laue technique, there are many metallurgical problems to which the Laue method is not applicable. To overcome the difficulties, Debye, Scherrer and Hull contrived an ingenious method for the investigation of samples available mostly in a pulverised form containing many hundreds or thousands of tiny, randomly orientated crystals. This technique is known as X-ray powder photography or simply the *powder method*.

In the powder method, the glancing angle θ is variable and the wavelength λ of the incident X-rays is kept constant. The variation of θ is provided by the multitude of differently orientated crystals, while the constancy of λ is maintained with appropriately filtered X-rays. The formation of the powder rings may be explained with the aid of Fig. 6.1. In this diagram, the incident beam IOC encounters a polycrystalline specimen where diffraction will occur if the normal ON of plane (hkl) intersects the reference sphere at $90°-\theta$ to I. The diffracted beam will then intersect the flat film at P. In a randomly orientated crystalline powder, there must be a great many small pulverised crystals with an (hkl) plane whose normal intersects the sphere at $90°-\theta$ to I; hence the points of intersection of these normals and the sphere will lie on the circle LMN. Similarly, the beams diffracted by these planes will intersect the sphere at points lying on the circle PQR. If a sufficient number of small crystals are present, the series of diffraction spots lying on the circle *PQR* will form a continuous ring.

In the same way, another set of planes ($h'k'l'$), whose spacings are different from those of the (hkl) planes, will satisfy diffraction conditions at a glancing angle θ'. The resultant diffraction will obey

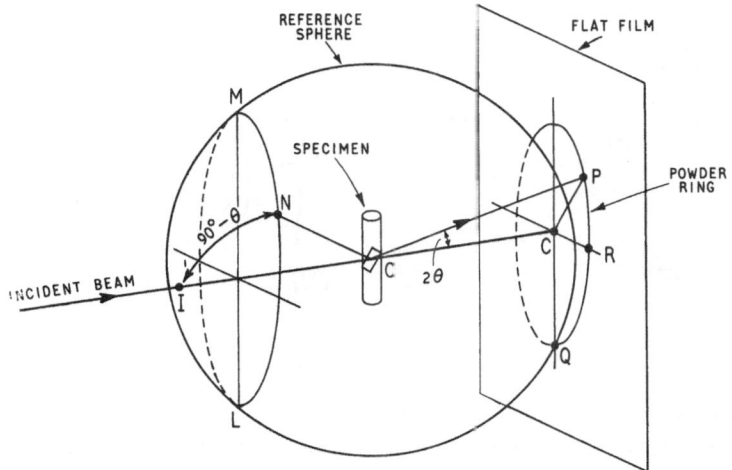

Fig. 6.1. Formation of a powder ring on a flat film

the same geometrical relationship as that shown in Fig. 6.1: the different glancing angle θ' (which is related to the different spacings of the atomic planes) will cause the diffraction ring to differ only in radius. Thus the radius of a diffracted powder line is directly related to the spacing of the set of atomic planes responsible for its existence. Since there is a large number of atomic planes in a space lattice, powder photographs show an appropriately large number of concentric rings; these constitute the *powder pattern*.

For practical purposes, the limitations of flat film recording soon become obvious. Because of the rapid increase in the radius of diffraction rings with increasing values of θ, only a few diffraction lines can be recorded on a flat plate. Even if the flat film is imagined as infinitely extended, it can only be used if the angle *POC* of Fig. 6.1 is less than 90°. Thus θ must be less than 45°; at any glancing angle greater than 45°, the diffracted X-rays cannot be recorded on the same flat film.

These limitations are overcome in a method devised by Debye

109

and Scherrer, and independently by Hull, in which the film is wrapped round a cylindrical surface concentric with the specimen, and the diffracted beams are allowed to fall upon the full circumference of the film. This experimental arrangement is shown

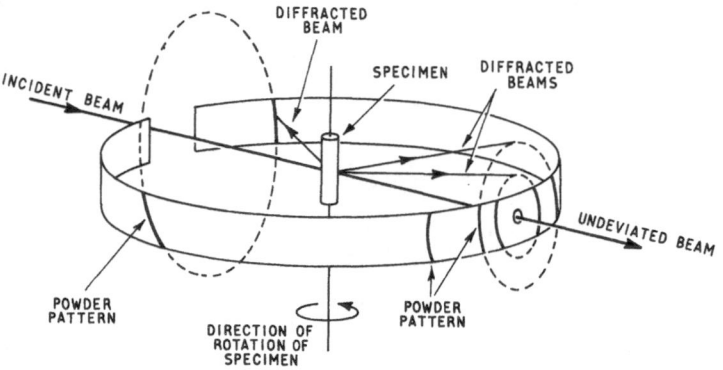

Fig. 6.2. Formation of a powder pattern on a cylindrical film

diagrammatically in Fig. 6.2. In order to bring a large number of small crystals into reflecting positions, the specimen is rotated, as shown in Fig. 6.2. The spottiness of some diffraction patterns is often the result of a stationary specimen, or of a specimen containing too few (or excessively large) crystals. However, as more crystals are brought into a reflecting position, the spottiness gradually disappears, blending into lines of even intensity on the film (see Fig. 6.4 and Fig. 6.5).

The basic apparatus needed for powder work is simple. However, because of the importance of the method, a great deal of thought has been given to the design and construction of cameras and of instruments for measuring line spacings. One of the essential properties of an efficient powder camera is repeatability of photographs — which means that it must be possible to position a specimen in the centre of the cylindrical film with extreme accuracy every time a photograph is taken. It is also convenient if the camera diameter bears some relationship to the atomic spacing, so that millimetre measurements taken from the flattened film can be simply converted to θ or 2θ values as required for further calculations.

Safety features and precautions involved in such instruments should also be considered. The undeviated X-ray beam must be

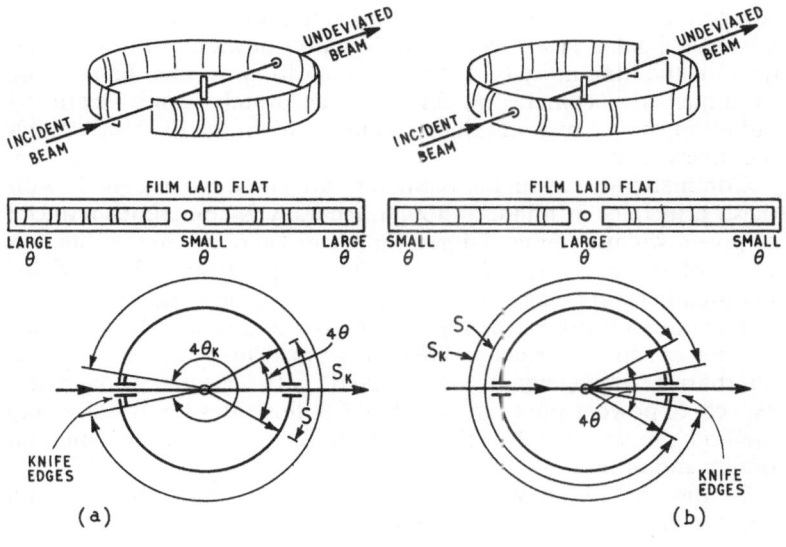

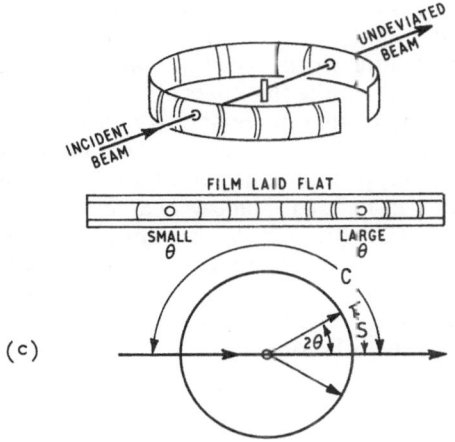

Fig. 6.3. Methods of film mounting in powder cameras: (a) Bradley–Jay mounting;
(b) van Arkel mounting; (c) Ievinš–Straumanis mounting

stopped in some way, so that the operator is fully protected; however, back-scatter, which would cause unwanted lines or fogging on the film, must be avoided. In addition, it must be possible to adjust the camera quickly and accurately, since scattered radiation caused by a misaligned camera might inadvertently reach the investigator.

Commercially available cameras are mainly designed with these principles in mind, although accuracy is sometimes reduced to give a slight commercial gain in price. There are many different types of cameras available, some being specifically designed for high-temperature work and others for cryogenic investigations.

For most purposes, the film can be placed around the specimen in one of three basically different ways. The reasons for these alternative arrangements are linked with certain geometrical aspects of powder photography. Fig. 6.3 illustrates the Bradley–Jay method, the van Arkel method and the Ieviņš–Straumanis method of film mounting.

In the method developed by Bradley and Jay, the glancing angle θ may be calculated from the expression

$$\theta = \frac{S}{4R} \tag{6.1}$$

where S is the diameter of the diffraction ring (i.e. the distance between corresponding diffraction lines), and R is the radius of the camera. This expression is sufficient for most applications, although for precision work some secondary effects must be considered — such as the finite thickness of the film and its shrinkage during processing. It is also necessary to consider manufacturing deviation in the mean camera diameter. To minimise this source of error, the film is slipped under a pair of knife edges which, under radiation, cast a well-defined shadow near each end of the film. In this way, a standard distance S_K is imprinted on the film, and this will shrink in the same proportion as the distance between corresponding diffraction lines. If the subtended angle $4\theta_K$ between the knife edges is known, the Bragg angle for any pair of lines can be calculated from the relationship:

$$\frac{\theta}{\theta_K} = \frac{S}{S_K} \tag{6.2}$$

The Bradley–Jay method of film mounting is very useful in investigations of lattice spacing at small values of θ. For large values of θ, the error in θ_K is proportionately higher and adversely affects the precision attainable.

The method of mounting suggested by van Arkel is essentially the reverse of the Bradley–Jay method. The relationship between S, S_K and θ can be found in the following way. If S is again the spacing between corresponding diffraction lines, then

$$S = 2\pi R - 4\theta R \tag{6.3}$$

where θ is the corresponding Bragg angle in radians. This can be written in the form:

$$S = 4R\left(\frac{\pi}{2} - \theta\right) \tag{6.4}$$

Similarly, for the knife edges,

$$S_K = 4R\left(\frac{\pi}{2} - \theta_K\right) \tag{6.5}$$

Expressing θ and θ_K in degrees gives

$$S\frac{180°}{\pi} = 4R(90° - \theta) = 4R\psi \tag{6.6}$$

where

$$\psi = 90° - \theta$$

and

$$S_K\frac{180°}{\pi} = 4R(90° - \theta_K) = 4R\psi_K \tag{6.7}$$

where

$$\psi_K = 90° - \theta_K$$

From Equations 6.6 and 6.7

$$\frac{S}{S_K} = \frac{\psi}{\psi_K}$$

or

$$S = \psi\,\frac{S_K}{\psi_K} \tag{6.8}$$

The ratio S_K/ψ_K is a characteristic constant of the camera, and is

113

usually measured by means of a powder of known lattice parameter. Alternatively, for less accurate work, the camera constant is calculated from Equation 6.7 as:

$$\frac{S_K}{\psi_K} = \frac{4\pi R}{180°} \tag{6.9}$$

For a standard 9 cm diameter camera, this constant is taken as 0·31415 or $\pi/10$. The advantage of the van Arkel mounting is that the lines corresponding to high values of θ are closer together, and the distance between them is therefore relatively little affected by uneven film shrinkage.

The third method, devised by Ieviņš and Straumanis, eliminates the need to calibrate the camera for high precision. Two holes are punched in the film, at a known distance C apart, to allow the unobstructed passage of the X-ray beam [as shown in Fig. 6.3(c)]. The angle θ may then be calculated from the expression:

$$\theta = \frac{\pi S}{2C} \tag{6.10}$$

This method combines the advantages of the two preceding methods, for the small additional expense of a rather more complicated camera.

In metallurgical investigations, the phase changes that occur at high temperatures must often be determined. The camera available from Unicam, for example, has been specially designed to operate from room temperature up to 1,400°C. It is suitable for powder and fibre specimens, rotated as in normal Debye–Scherrer cameras, and also for small block specimens.

Low-temperature cameras are also needed for many studies of fatigue or for low-temperature brittle phase determinations. The usual method of maintaining a low temperature is to allow a narrow stream of liquid gas to trickle over the specimen throughout the exposure period. Diffraction caused by the cooling agent will only appear as a diffuse halo near the centre of the film, while the incoherent scattering from the liquid will slightly increase the background blackening of the photographic plate.

6.2 DETERMINATION OF UNIT CELL DIMENSIONS BY ANALYTICAL METHODS

Seven examples of X-ray photographs, all taken with copper $K\alpha$ radiation, are shown in Fig. 6.4. The van Arkel method of mounting was used for each of these films. It can be seen from the photographs that the number of lines differs from one pattern to the other. The reason for these variations has already been shown in Section 4.3: depending on the lattice type, some powder lines corresponding to certain lattice spacings are forbidden in accordance with the structure factor equation. The previous findings for reflection conditions can be summarised as follows:

Unit cell type	Reflection condition
Primitive	none
Face centred	h, k and l all even or all odd
Body centred	$h+k+l$ even
Hexagonal	$\begin{cases} h+2k = 3n, l = 2n \\ h+2k = 3n\pm1, l = 2n+1 \\ h+2k = 3n\pm1, l = 2n \end{cases}$

The determination of the unit cell dimension of a cubic substance is largely simplified by the proportionality which exists between the spacings of atomic planes. The lattice spacing in a cubic cell is given by the equation

$$d = \frac{a}{(h^2 + k^2 + l^2)^{\frac{1}{2}}} \qquad (6.11)$$

where a is the unit cell dimension, and d is the spacing of planes whose indices are hkl. In a cubic pattern, therefore, the lines are spaced in the ratios $1/\sqrt{2}$, $1/\sqrt{3}$, $1/\sqrt{4}$, etc., except when reflection conditions forbid their presence. The reversal of this rule indicates that, whenever the line spacing shows the above proportionality, the substance is evidently cubic. With some practice it is possible to recognise the cubic pattern by simple inspection.

The unit cell dimensions are usually evaluated in two steps. First, all the lines of the powder pattern are indexed, so that the relationship between the lattice planes and the line spacings can be found. Second, from the known lattice spacings, the unit cell parameters can be deduced. The following example shows how this principle can be applied to the determination of the unit cell dimensions of silver [see Fig. 6.4(d)]; a suitable layout for indexing the powder lines is shown in Table 6.1.

The lines of the powder pattern are first numbered from left to right for identification purposes, and their spacings are then measured by means of a travelling microscope (each line spacing S is found by deducting the left-hand reading from the corresponding right-hand reading). Next, the line spacings are divided by the

115

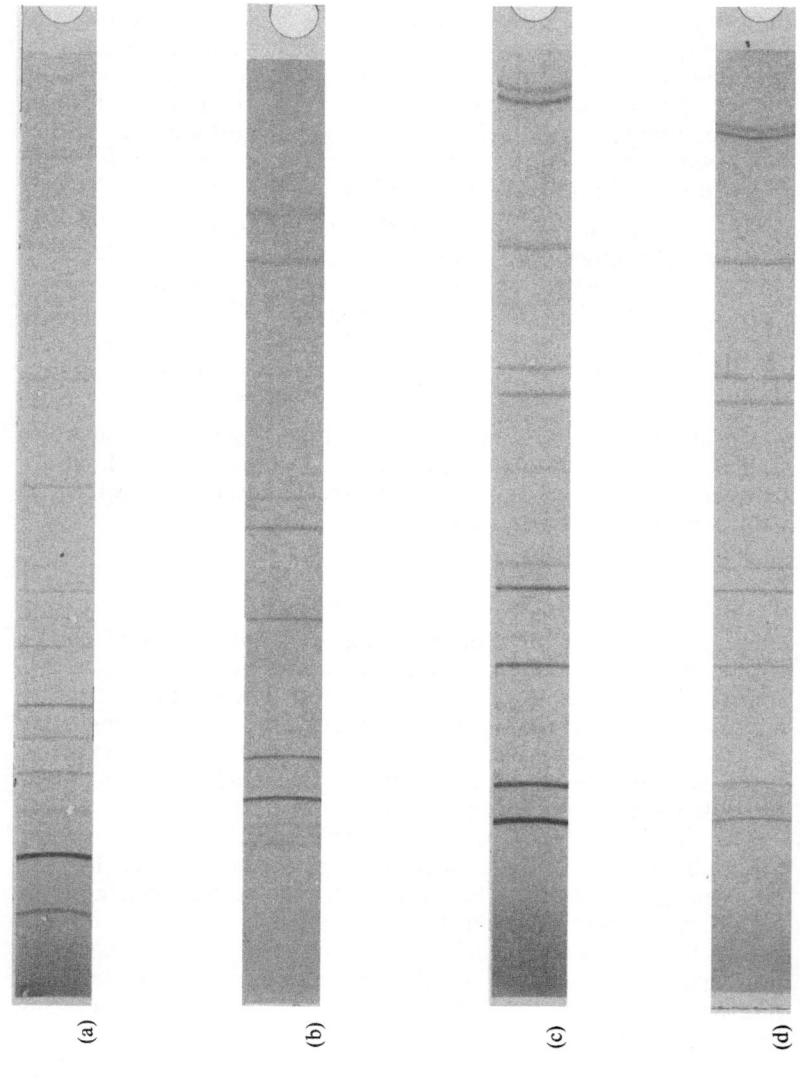

(a)

(b)

(c)

(d)

Fig. 6.4. Powder photographs of some elements and compounds (half of each photograph only is shown): (a) ammonium chloride (primitive cubic, a = 3·9495 Å); (b) copper (FCC, a = 3·6153 Å); (c) aluminium (FCC, a = 4·0490 Å); (d) silver (FCC, a = 4·0854 Å); (e) tungsten (BCC, a = 3·1648 Å); (f) magnesium (CPH, a = 3·2092 Å, c = 5·2103 Å); (g) tungsten carbide (CPH, a = 2·91 Å, c = 2·82 Å)

(e)

(f)

(g)

camera constant to give the ψ values (Equation 6.8). The Bragg angles are obtained by deducting each value of ψ from 90°, and each $\sin^2\theta$ is then evaluated by means of trigonometrical tables. The next step is to examine the data so far calculated, and to decide whether the unit cell is cubic or whether it belongs to a less symmetrical system. The simplest method is to establish whether the $\sin^2\theta$ values contain a common factor.

Table 6.1.

Layout for Indexing the Powder Lines of Silver

Line	Left-hand side (LH)	Right-hand side (RH)	$S = RH - LH$	$\psi = \dfrac{S}{S_K/\psi_K}$	$\theta = 90° - \psi$	$\sin^2\theta$	N	Indices of reflection
1	5·820	28·095	22·275	70·73°	19·27°	0·1090	3	111
2	6·310	27·605	21·295	67·72°	22·28°	0·1438	4	200
3	7·900	26·020	18·120	57·69°	32·31°	0·2857	8	220
4	8·235	25·675	17·440	55·39°	34·61°	0·3226	9	{300 221
5	8·915	25·000	16·085	51·19°	38·81°	0·3928	11	311
6	9·245	24·675	15·430	49·11°	40·89°	0·4286	12	222
7	10·525	23·490	12·965	41·28°	48·72°	0·5648	16	400
8	11·530	22·385	10·855	34·55°	55·45°	0·6784	19	331
9	11·880	22·045	10·165	32·35°	57·65°	0·7137	20	420
$10\alpha_1$	13·425	20·495	7·070	22·46°	67·54°	0·8541	24	422
$10\alpha_2$	13·485	20·425	6·940	22·10°	67·90°	0·8585	24	422
$11\alpha_1$	15·120	18·785	3·665	11·67°	78·33°	0·9591	27	{511 333
$11\alpha_2$	15·260	18·660	3·400	10·82°	79·18°	0·9647	27	{511 333

The existence of this common factor can be shown by equating the interatomic spacing d calculated from Equation 6.11 and the value d expressed by the Bragg condition. Thus:

$$\frac{\lambda}{2\sin\theta} = \frac{a}{(h^2+k^2+l^2)^{\frac{1}{2}}} \tag{6.12}$$

Taking $N = h^2 + k^2 + l^2$ gives:

$$N = \frac{4a^2\sin^2\theta}{\lambda^2} \tag{6.13}$$

In Equation 6.13, the ratio $4a^2/\lambda^2$ is constant (since the lattice parameter a and the wavelength λ are constants) and N must be

118

an integer (since h, k and l are integers). It should thus be a simple matter of a few trials to find the common factor of the $\sin^2 \theta$ values.

The first three $\sin^2 \theta$ values for silver were found to be 0·1090, 0·1438 and 0·2857, and the common factor of these values is approximately 0·0358. If each $\sin^2 \theta$ value is now divided by this factor, the values of N can be obtained. If the resultant figures are close to integers (in this example 3, 4 and 8 for the first three lines), the substance is cubic.

A few trials will show how to assign the correct indices to each line. For example, when $N = 3$, the only possible indices are 111; and when $N = 4$, the possible hkl indices are 200. Indices for all other values of N can be found by similar reasoning. Since the indices are unmixed (i.e. they are all odd or all even), it is at once clear that the cubic lattice is face centred. In fact the fourth line (for which $N = 9$) is found to be an exception, since the indices could be either 300 or 221. This spurious line probably belongs to a very strongly reflecting plane of some impurity contained in the silver. In further calculations, this line should be ignored.

The final stage of the procedure is to find the lattice parameter a from the data already calculated. From Equation 6.13, it is evident that the common factor present in all the $\sin^2 \theta$ values is $\lambda^2/4a^2$. Therefore, in this example,

$$\frac{\lambda^2}{4a^2} = 0.0358$$

Therefore

$$a = \left(\frac{\lambda^2}{4 \times 0.0358} \right)^{\frac{1}{2}} = \left[\frac{(1.5418)^2}{4 \times 0.0358} \right]^{\frac{1}{2}} = 4.0755 \text{ Å}$$

Precision measurements on silver have shown that the actual lattice parameter a is 4·0854 Å at 20°C. Thus the error involved in the rapid method of calculation used above is about $\frac{1}{4}\%$.

Further refinement can be achieved by finding the average of the values of a calculated for each line with the basic equation

$$a^2 = \frac{N\lambda^2}{4 \sin^2 \theta}$$

This method is illustrated in Table 6.2, from which the average value of the lattice parameter a for silver was found to be 4·0796 Å. This is obviously closer to the generally accepted value of 4·0854 Å.

Consider now the Bragg equation in the form:

$$d \sin \theta = \frac{n\lambda}{2} \tag{6.14}$$

Differentiating both sides gives

$$d \cos \theta \, \Delta\theta + \sin \theta \, \Delta d = 0 \tag{6.15}$$

and rearranging Equation 6.15 then gives

$$\frac{\Delta d}{d} = -\cot \theta \, \Delta\theta \tag{6.16}$$

The proportional error in the measurement of the line spacing is given by the expression $\Delta d/d$, and $\cot \theta \, \Delta\theta$ is the associated angular error. When θ tends to $90°$, $\cot \theta$ tends to zero; therefore, from

Table 6.2.

Calculation of the Lattice Parameter for the Powder Lines of Silver

Line	Indices of reflection	$N = (h^2 + k^2 + l^2)^{\frac{1}{2}}$	$\dfrac{N}{\sin^2\theta}$	a^2	$a\,(\text{Å})$
1	111	3	27·524	16·331	4·0449
2	200	4	27·823	16·539	4·0663
3	220	8	28·004	16·646	4·0800
4	$\left\{\begin{array}{l}300\\221\end{array}\right.$	9	—	—	—
5	311	11	28·003	16·646	4·0800
6	222	12	28·003	16·646	4·0800
7	400	16	28·327	16·838	4·1034
8	331	19	28·004	16·647	4·0803
9	420	20	28·035	16·665	4·0823
$10\alpha_1$	422	24	28·094	16·669	4·0827
$10\alpha_2$	422	24	27·952	16·665	4·0823
$11\alpha_1$	$\left\{\begin{array}{l}511\\333\end{array}\right.$	27	28·158	16·707	4·0874
$11\alpha_2$	$\left\{\begin{array}{l}511\\333\end{array}\right.$	27	27·990	16·888	4·0851

Equation 6.16, any errors in the value of d (and consequently in the value of a) caused by errors in θ are proportionately smaller as θ approaches $90°$. In the present example for silver, the lattice parameter a extrapolated to $\theta = 90°$ is 4·0865 Å.

The accuracy of the determination of lattice parameters is also affected by the divergence of the incident X-ray beam and by the absorption in the specimen. These two sources of error are usually treated together, although their physical effects are quite distinct. It has been shown experimentally that, for general powder work, the error caused by absorption and divergence is proportional to:

$$\frac{1}{2}\left(\frac{\cos^2\theta}{\sin^2\theta} - \frac{\cos^2\theta}{\theta}\right)$$

A more accurate determination of the lattice parameter can be found if successive values of a are plotted against $\frac{1}{2}(\cos^2\theta/\sin^2\theta + \cos^2\theta/\theta)$ and are then extrapolated to zero by means of the best straight line drawn through the points.

All the above methods of increasing the accuracy of lattice parameter determinations are only concerned with diffractive errors and errors of line spacing measurements. Naturally, some variations in the manufacture of cameras must also be allowed for: thus the value of the camera constant may vary for similarly constructed cameras. Such a discrepancy may well have affected the results in the above calculations for silver, since a nominal constant of $\pi/10$ for a 9 cm diameter camera was used throughout. A camera constant can be measured accurately if the camera is used to produce a powder photograph of a known simple cubic substance. The line spacings are measured in the usual way. Since the lattice parameter of the cubic element is known, the lines can be indexed by little more than simple inspection of the pattern. The values of $\sin^2\theta$ can then be calculated from Equation 6.13. Each value of ψ is obtained by deducting θ from $90°$, and the experimental camera constant is then given by the ratio S/ψ (which, from Equation 6.8, is equal to S_K/ψ_K). Again, for higher accuracy, the camera constants should be extrapolated to $\theta = 90°$.

Fig. 6.4(f) is a powder photograph of magnesium. If indexing of these powder lines is attempted by the method outlined above, some very confusing results will be found — indicating that this substance belongs to some other crystal system. When unknown lattice parameters are to be determined, it is helpful to remember that the large majority of metals belong to either the cubic or the hexagonal system. Considerable time and effort are saved if it is recognised that the metal is not cubic but hexagonal, and that a modified indexing process should therefore be followed.

The principles of indexing the pattern logically are the same as for the cubic system: namely to establish whether there is a

121

particular relationship between the $\sin^2\theta$ values. The plane spacing d in a hexagonal lattice is given by the equation:

$$\frac{1}{d^2} = \frac{4}{3}\left(\frac{h^2 + hk + k^2}{a^2}\right) + \frac{l^2}{c^2} \tag{6.17}$$

The reciprocal of the square of the plane spacing may also be expressed from the Bragg condition as:

$$\frac{1}{d^2} = \frac{4\sin^2\theta}{\lambda^2} \tag{6.18}$$

Equating Equations 6.17 and 6.18 gives

$$\frac{4\sin^2\theta}{\lambda^2} = \frac{4}{3}\left(\frac{h^2 + hk + k^2}{a^2}\right) + \frac{l^2}{c^2} \tag{6.19}$$

or

$$\sin^2\theta = A(h^2 + hk + k^2) + Bl^2 \tag{6.20}$$

where

$$A = \frac{\lambda^2}{3a^2}$$

and

$$B = \frac{\lambda^2}{4c^2}$$

Again it can be seen that the $\sin^2\theta$ values contain common factors A and B. The problem now is to find these values; the calculations can be considerably simplified if the first trials are restricted to the $hk0$ reflections, thus eliminating the factor B:

$$\sin^2\theta_{100} = A \qquad\qquad \sin^2\theta_{210} = 7A$$

$$\sin^2\theta_{110} = 3A \qquad\qquad \sin^2\theta_{300} = 9A$$

$$\sin^2\theta_{200} = 4A \qquad\qquad \sin^2\theta_{220} = 12A$$

It is significant that the ratio 3 occurs frequently ($3A/A$, $9A/3A$, $12A/4A$, etc.). This ratio is in fact representative of the hexagonal system, and it cannot occur in other systems except by chance. If the ratio of the $\sin^2\theta$ values of two of the first three or four lines is 3, the first one is probably the 100 line, and the other the 110 line. By chance, or for structural reasons, the 100 reflection may be absent; the lines may then be the 110 and 300 reflections.

122

Once the value of A has been found, the other lines can be indexed and the value of B determined. The lattice parameters can be calculated by applying the same principles as for cubic systems.

Table 6.3 shows the line spacings and the $\sin^2\theta$ values for the first nine lines of magnesium (the camera constant again being taken as 0·31415). The next step is to look for the significant ratio of 3 between two $\sin^2\theta$ values. The ratio of the second $\sin^2\theta$ value

Table 6.3.

Layout for the Determination of the $\sin^2\theta$ Values of Magnesium

Line	S	$\psi = \dfrac{S}{S_K/\psi_K}$	$\theta = 90° - \psi$	$\sin^2\theta$
1	25·575	81·380	0·620	0·0225
2	23·285	74·120	15·880	0·0748
3	22·930	72·988	17·012	0·0856
4	22·512	71·661	18·339	0·0990
5	20·815	66·258	23·742	0·1621
6	19·890	63·312	26·588	0·2018
7	19·215	61·155	28·345	0·2326
8	18·610	59·238	30·762	0·2617
9	18·425	58·649	31·351	0·2707

Table 6.4.

Layout for the Determination of the Constant B for Magnesium

Line	$\sin^2\theta$	$\sin^2\theta - A$	$\sin^2\theta - 3A$
1	0·0225		
2	0·0748		
3	0·0856	0·0106	
4	0·0990	0·0240	
5	0·1621	0·0871	
6	0·2018	0·1268	
7	0·2326	0·1576	0·0076
8	0·2617	0·1867	0·0367
9	0·2707	0·1957	0·0457

to the first $\sin^2\theta$ value is close to 3 (it is actually 3·33); also the ratio of the seventh to the second $\sin^2\theta$ value is very close to 3 (in fact it is 3·11). As the ratio 3·11 is closer to 3 than 3·33, the first assumption will be that the second line is the 100 reflection and the

123

seventh line is the 110 reflection. The real test of whether this assumption is correct is the success with which all the other lines can be indexed to give a reasonable correlation between measured and calculated $\sin^2\theta$ values.

If the l index of an hkl line is not zero, then the $\sin^2\theta$ value of that line must contain a term Bl^2, in accordance with Equation 6.20. To facilitate the search for the constant B, the characteristic multiples of A will be subtracted from the $\sin^2\theta$ values, as shown in Table 6.4 If A is taken as 0·0750 (i.e. 0·0748 corrected to two significant figures), then $3A = 0·2250$. It can be seen that the value

Table 6.5.

Determination of the Indices of Reflection of the
Powder Lines of Magnesium

Line	Indices of reflection	$\sin^2\theta$ Experimental	$\sin^2\theta$ Calculated	Error
1	001	0·0225	0·0232	0·0007
2	100	0·0748	0·0750	0·0002
3	002	0·0856	0·0928	0·0072
4	101	0·0990	0·0982	0·0008
5	102	0·1621	0·1678	0·0057
6	003	0·2018	0·2088	0·0070
7	110	0·2326	0·2250	0·0076
8	111	0·2617	0·2482	0·0135
9	103	0·2707	0·2838	0·0131

$0·0232 \pm 0·0008$ is present in the first and fourth lines, and may therefore correspond to the desired value of B when $l = 1$. It will now be assumed that the constant $A = 0·0750$ and that $B = 0·0232$; these values will be used to index the lines. At this stage the indexing must be carried out with extreme caution because the values of A and B have been derived only from the low-angle reflections, at which spacing errors are a maximum (page 114).

From Equation 6.20, $\sin^2\theta_{001} = B$; therefore the first line is the reflection of the (001) plane. The second line has already been taken as a 100 reflection. The third line is most probably an 002 reflection, since $\sin^2\theta_{002} = Bl^2 = 0·0928$. To a first approximation, the value 0·0856 may be taken as within experimental error. Similarly, the fourth line is a 101 reflection, for which the calculated $\sin^2\theta$ value is 0·0982. In the same manner all nine lines may be indexed: the results obtained are shown in Table 6.5.

The lattice parameters can be found from the values of A and B, since

$$A = \frac{\lambda^2}{3a^2}$$

and

$$B = \frac{\lambda^2}{4c^2}$$

Substitution of A and B and of the wavelength of copper $K\alpha$ radiation into these relationships gives $a = 3 \cdot 2504$ Å and $c = 5 \cdot 0612$ Å. Precision measurements give these parameters as $a = 3 \cdot 2092$ Å and $c = 5 \cdot 2103$ Å. The difference between the experimental results calculated above and the precision measurements can be considerably reduced if the more refined techniques, discussed in connection with the work on cubic silver, are used. In general, the highest accuracy obtainable in parameter measurements from powder patterns is about $0 \cdot 02 \%$; this can be achieved with high-angle doublets which are sharp and well resolved. When high-angle doublets are slightly broadened, accuracy gradually diminishes to about $0 \cdot 2 \%$. If high-angle reflections are blurred (as often happens with less symmetrical systems) and only low-angle reflections can be used, an accuracy between $0 \cdot 1 \%$ and 1% is all that can be expected.

In the powder photographs of sodium chloride (FCC, $a = 5 \cdot 639$ Å) shown in Fig. 6.5, some pairs of lines can be seen at the high-angle end of the spectra. These doublets are formed by the $K\alpha_1$ and $K\alpha_2$ components of the characteristic $K\alpha$ radiation. The use of high-angle doublets is extremely important in precision work: not only are the spacing errors at a minimum, but the wavelengths of $K\alpha_1$ and $K\alpha_2$ radiation are known to within $0 \cdot 001 \%$.

At the end of a spectrum where the low-angle reflections are recorded, some line pairing can occasionally be found. However, this is not caused by the resolution of the $K\alpha$ components; it is a result of the fact that, in forward diffraction, the diffracted rays have slightly different path lengths within the cylindrical specimen, and hence different amounts of radiation are absorbed. Some rays pass near the centre of the specimen and are almost completely absorbed, while others travelling along a chord of the cylinder suffer less absorption. The photographic result of this difference is that one line of the pair appears darker than the other, while the area between the lines is considerably lighter than either line.

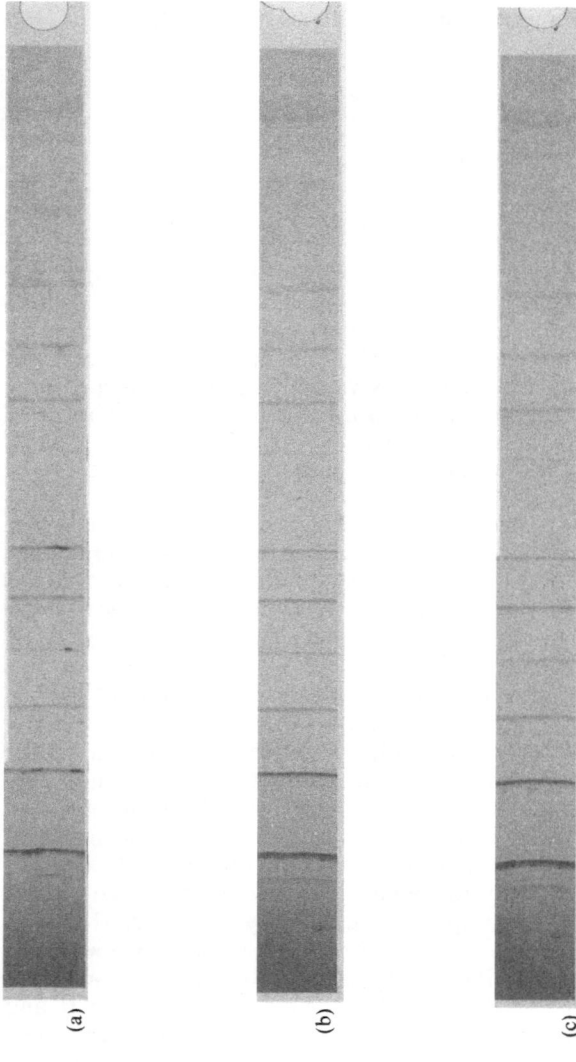

Fig. 6.5. Powder photographs of sodium chloride (half of each photograph only is shown): (a) stationary specimen; (b) rotating specimen, but uneven particle size; (c) normal powder pattern

Substances belonging to the tetragonal system may be indexed in a manner similar to that used for the hexagonal magnesium, the appropriate equation being

$$\sin^2\theta = C(h^2 + k^2) + Dl^2 \qquad (6.21)$$

where $C = \lambda^2/a^2$ and $D = \lambda^2/c^2$. If the possible combinations of h and k are now considered, with l being zero, the following relationships are obtained:

$$\sin^2\theta_{100} = C \qquad \sin^2\theta_{210} = 5C$$
$$\sin^2\theta_{110} = 2C \qquad \sin^2\theta_{220} = 8C$$
$$\sin^2\theta_{200} = 4C \qquad \sin^2\theta_{300} = 9C$$

The ratio of 2 which applies to the $\sin^2\theta$ values of the 100, 110, 200 and 220 reflections is characteristic of the tetragonal system.

The direct analytical method is much more difficult and troublesome for the biaxial systems (orthorhombic, monoclinic and triclinic), particularly when the unit cell size is large. However, some analytical methods based on the reciprocal lattice concept are available, but these are beyond the scope of this book.

6.3 DETERMINATION OF UNIT CELL DIMENSIONS BY GRAPHICAL METHODS

It must now be clear that indexing any but the simplest of cubic patterns is extremely laborious and full of pitfalls to trap the beginner. It is not unusual to spend most of a working day trying to index a pattern of unknown origin, only to find eventually that it belongs to some mixture of two or more cubic elements.

To overcome some of the hard work involved in obtaining even approximate lattice parameters, many graphical aids have been designed in the past. The simplest of these graphical devices is for the cubic system. However, the application of charts to the less symmetrical uniaxial systems (tetragonal, hexagonal and trigonal) is infinitely more useful.

The interplanar spacing d and the hkl indices of a cubic system are related by the equation

$$d = \frac{a}{(h^2 + k^2 + l^2)^{\frac{1}{2}}} \qquad (6.22)$$

Since h, k and l are integers and are the same for all cubic substances, the interplanar spacing d is proportional to the lattice parameter a only; the spacings of cubic crystals differ in scale only.

Fig. 6.6 shows one suitable chart for indexing cubic patterns. The values of d must first be calculated from the measured $\sin \theta$ values, and then plotted on a paper strip along the same scale as in

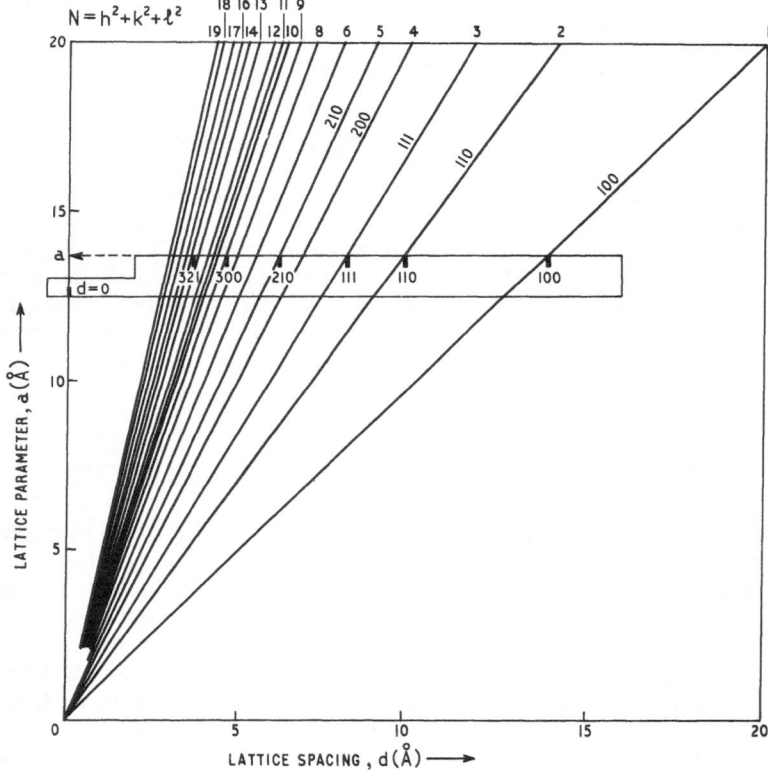

Fig. 6.6. Indexing chart for cubic patterns

the chart. The paper strip is next translated parallel to the horizontal axis of the chart, the $d = 0$ point being kept on the vertical axis. When the markings on the paper strip coincide with the lines of the chart, the indices of each line are simply read off the chart; an approximate lattice parameter a is given by the vertical location of the strip.

128

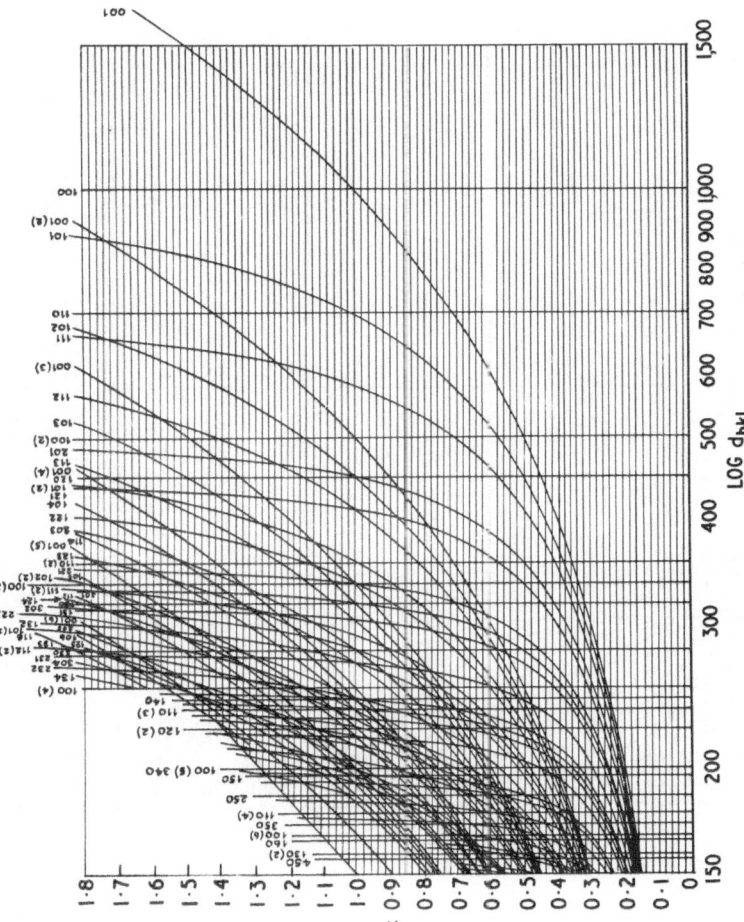

Fig. 6.7. The Hull–Davey chart for tetragonal crystals

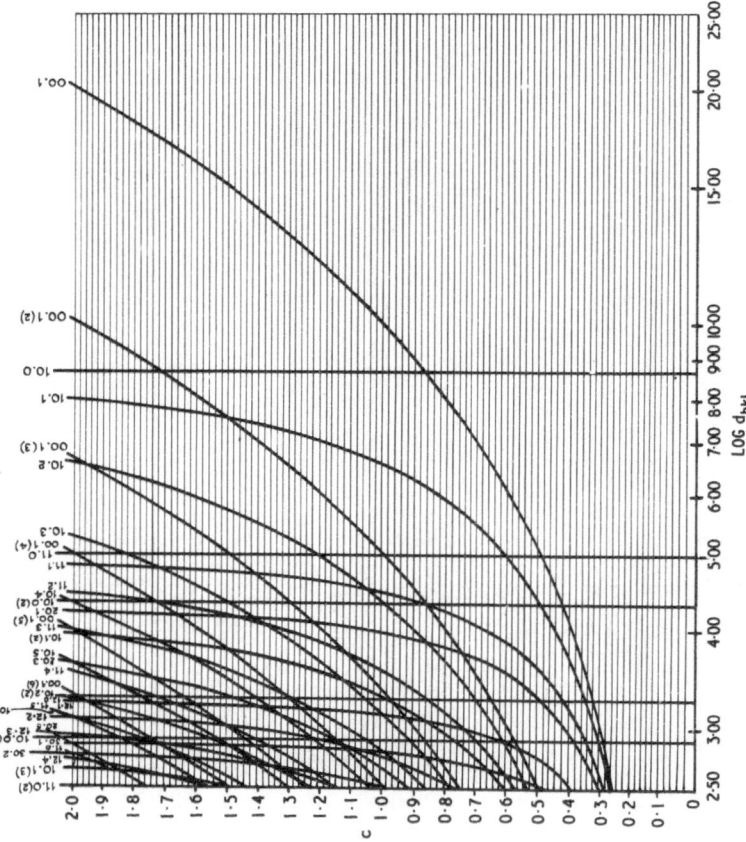

Fig. 6.8. The Hull–Davey chart for hexagonal crystals

In the tetragonal system, the interplanar spacing d is related to the lattice parameters a and c by the equation

$$d = \frac{1}{[(h^2 + k^2)/a^2 + l^2/c^2]^{\frac{1}{2}}} \tag{6.23}$$

If Equation 6.23 is squared, and the logarithm taken of each side, then

$$2\log d = 2\log a - \log(h^2 + k^2 + l^2/E^2) \tag{6.24}$$

where $E^2 = c^2/a^2$.

Similarly, in the hexagonal or trigonal systems, lattice spacings are given by the equation:

$$d = \frac{1}{[\frac{4}{3}(h^2 + hk + k^2)/a^2 + l^2/c^2]^{\frac{1}{2}}} \tag{6.25}$$

Again squaring both sides of Equation 6.25 and taking logarithms gives

$$2\log d = 2\log a - \log\left[\frac{4}{3}(h^2 + hk + k^2) + l^2/E^2\right] \tag{6.26}$$

where $E^2 = c^2/a^2$.

Hull and Davey used Equations 6.24 and 6.26 as the basis of charts for the graphical indexing of tetragonal and hexagonal powder patterns. The standard Hull–Davey charts are shown in Fig. 6.7 and Fig. 6.8, and their use will now be illustrated with an actual example.

Table 6.6.

Layout for the Determination of the d Values of Tungsten Carbide

Line	Left-hand side (LH)	Right-hand side (RH)	$S = RH - LH$	$\psi = \dfrac{S}{S_K/\psi_K}$	$\theta = 90° - \psi$	$\sin^2\theta$	$\dfrac{1}{d^2}$	d
1	3·895	27·300	23·405	74·508°	15·492°	0·0713	0·1195	2·893
2	4·220	26·980	22·760	72·444°	17·556°	0·0910	0·1530	2·556
3	4·845	26·345	21·500	68·437°	21·563°	0·1350	0·227	2·099
4	5·225	25·975	20·750	66·050°	23·950°	0·1647	0·276	1·903
5	6·475	24·725	18·250	58·092°	31·908°	0·2793	0·469	1·460
6	6·615	24·590	17·975	57·217°	32·783°	0·2932	0·493	1·424
7	7·195	24·005	16·810	53·509°	36·491°	0·3536	0·594	1·297
8	7·385	23·825	16·440	52·330°	37·670°	0·3732	0·626	1·264
9	7·515	23·685	16·170	51·471°	38·529°	0·3880	0·652	1·238
10	8·065	23·135	15·070	47·969°	42·031°	0·4482	0·753	1·152
11	9·215	21·980	12·765	40·633°	49·367°	0·5760	0·970	1·015

In Table 6.6, the readings obtained from the powder photograph of tungsten carbide are listed. First the $\sin^2\theta$ values were calculated and then, with the help of the quadratic form of the Bragg equation, the d values were found. These d values must be plotted on a strip along the same logarithmic scale as in the chart. In order to do this, the paper strip is placed against the horizontal logarithmic scale printed on the chart and is marked at the appropriate points, as illustrated in Fig. 6.9. The paper strip is then placed on the chart

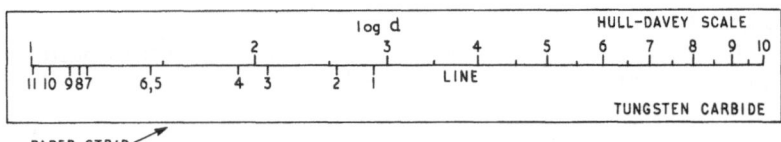

Fig. 6.9. Indexing strip for tungsten carbide

and moved about — while being kept parallel with the horizontal axis. The vertical and horizontal movements represent a series of searches, trying different c/a ratios and a values. An acceptable match occurs when every line on the strip corresponds to a line on the chart. In the present example, this match is reasonable for $c/a = 0.97$. It should be noticed that it is not possible to match the first four lines of this example simultaneously; it must therefore be concluded that the third line is some spurious line belonging to an impurity in the tungsten carbide sample. The indices of lines are then found to be as shown in Table 6.7.

The lattice parameters can now be determined by substitution of d^2_{hk0} and d^2_{00l} into Equation 6.17, together with the appropriate $hk0$ and $00l$ indices. For example, for the 001 reflection, $1/d^2 = 0.1195$; substitution into Equation 6.17 gives $c = 2.89$ Å. Similarly, for the 100 reflection, $1/d^2 = 2.556$ and a then equals 2.94 Å. The lattice parameters of tungsten carbide found by precision analytical

Table 6.7.

Indices of Lines obtained from the
Powder Photograph of Tungsten Carbide

Line	Indices	Line	Indices	Line	Indices
1	001	5	110	9	102
2	100	6	002	10	201
3	–	7	111	11	112
4	101	8	200		

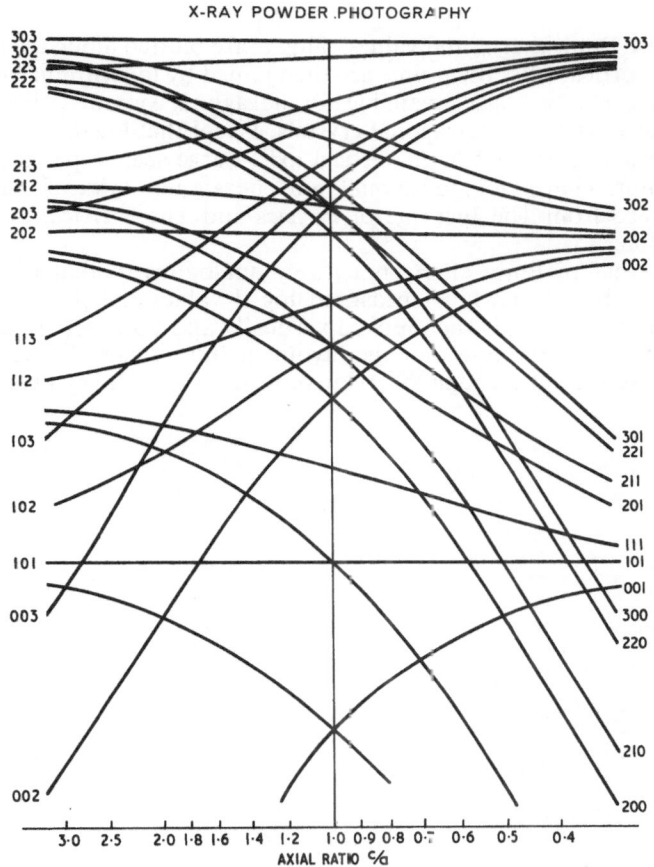

Fig. 6.10. The Bunn chart for tetragonal crystals

methods (and extrapolation of $\sin \theta$ to zero) are $c = 2.84$ Å and $a = 2.91$ Å. Comparison of these values with those obtained by graphical techniques shows that the considerable labour-saving value of graphical indexing is at the expense of about 1% loss in accuracy. This accuracy can be increased by simple calculations which will eliminate systematic errors.

For some purposes, particularly when the c/a ratio is small, the Hull–Davey charts are of limited use only. Instead, the Bunn charts (as illustrated in Fig. 6.10) can be used for tetragonal crystals. The values of $\log (h^2 + k^2)$ are plotted on the left-hand vertical axis

133

of these Bunn charts, and values of log l^2 are plotted along the right-hand vertical axis; the points are then joined by logarithmic curves. Bunn charts are also available for hexagonal systems; these are similar to those for tetragonal crystals, except that log $(h^2 + hk + k^2)$ is plotted on the left-hand vertical axis. Large-scale reproductions of Bunn charts for tetragonal and hexagonal systems can be obtained from The Institute of Physics and The Physical Society, London.

The use of Bunn charts is entirely analogous to that of Hull–Davey charts, in which measured log d values are plotted on a paper strip which is then moved vertically and horizontally until a match is found. Note that the layout of Bunn charts is such that they are turned through 90° relative to the Hull–Davey charts.

7

Applications of X-ray powder photography

7.1 IDENTIFICATION OF ELEMENTS AND COMPOUNDS

Undoubtedly, the widest application of powder photography today is the identification of compounds — either singly or in mixtures. It has already been shown that the line spacings (and hence the corresponding d values) of a powder pattern depend only on the unit cell dimensions of the substance, while the intensity distribution of the lines is affected by the relative positions of atoms within the unit cell. Any combination of these two factors — the unit cell dimensions and the atomic arrangement — is unique to a particular element or compound; hence their recognition can lead directly to the identification of a substance.

The complete determination, from a powder pattern, of the positions of all the atoms in a unit cell would be a very lengthy process. To avoid this, powder photographs are usually compared to standard powder patterns of known substances. If the unknown and standard powder patterns are identical, the substances giving rise to the patterns must also be identical. This principle eliminates the tedium of cell measurements and structure investigation; it is particularly useful for determining which members of an iso-structural group of compounds are present in a given specimen. For such a determination, a set of powder photographs of each member of the group must be prepared under standardised experimental conditions, and the powder pattern of the unknown sample must then be compared to the standard set. However, this procedure often becomes unmanageable if nothing at all is known about the sample: the number of standard photographs would have to be between about 2,000 and 5,000 to provide a close enough net, and consequently the process of comparison would require a great deal of time.

A very logical modification of the above procedure was proposed by Hanawalt and co-workers in the United States, and independently by Boldyrev and his colleagues in the Soviet Union. They suggested that, instead of standard films, a set of cards should be prepared bearing all the pertinent X-ray and optical characteristics of each substance. The classification of such cards is based on the d values of the three most intense lines. The intensities of the lines are measured on a relative scale: the darkest line is allotted an intensity of 100, and the intensity of each of the other two lines is visually estimated as a percentage of this. The data contained in the three d values and the relative intensities of the three lines are usually enough to reduce the nearly infinite possibilities to only a few probable compounds.

To compile the large amount of data required for these cards, British, American and Soviet institutions have pooled their resources and co-operated in publishing their findings. The body responsible for the publication of up-to-date data is the American Society for Testing Materials (ASTM). The present system is known as the ASTM X-ray powder data file, and consists of a card file system and an accompanying 'powder data index'.

If an unknown pattern is to be identified, the d values of its lines must first be listed in decreasing order of intensity. The d value of the most intense line is used to determine the Hanawalt group number in the index system. The d value of the second most intense line should then be matched against the second column of d values in the index, and so on for the third line. Once the closest correspondence between the experimental and indexed d values has been found, the relative intensities of the lines must be compared. Lastly, when good agreement has been achieved, the appropriate card is taken from the file and all the d values and relative intensity values are compared to the observed lines. Identification is complete when full agreement is obtained.

Sometimes, no proper match can be found between the observed and tabulated data. One obvious reason might be that the substance is not included in the file: this is still possible, even with well over 5,000 cards in the system. Alternatively, the indexed d value and the measured d value of the most intense line, which determines the Hanawalt number of the substance, may differ slightly. The appropriate correspondence between tabulated and observed values might then be found in one of the two adjacent Hanawalt groups.

The process of identification is relatively straightforward when the unknown substance is a single-phase compound. However, when mixtures of compounds are present, complications arise because the observed powder pattern is then made up of two or more superimposed individual patterns. In these circumstances, the five most intense lines of the pattern should be considered; different combinations of three out of these five d values can then be matched against the indexed values – remembering that the three most intense lines may not all be due to the same compound.

If the powder data file is to be used effectively for mixtures, experience, coupled with some intelligent guess-work, is required. Additional hazards are presented by chance coincidences: for example, two weak lines may be superimposed on the photograph, resulting in faulty estimates of relative intensities. Difficulties of this nature can often be overcome by means of some other method of qualitative analysis – however incomplete the results it may yield. It should be recognised that X-ray powder photography is only one of many methods of identification available to the experimenter. Techniques such as optical spectroscopy or X-ray fluorescence analysis may provide some vital information which enables mixture identification to proceed. On many occasions, for instance, spectroscopy and X-ray diffraction techniques have been used in conjunction for identifying the elements in a mixture, and for determining to what extent and in what form each of the elements is present.

7.2 ANALYSIS OF BINARY AND TERNARY THERMAL EQUILIBRIUM DIAGRAMS

Metals are sometimes completely soluble in each of the solid, liquid and vapour states. Much more frequently, however, their solubilities are restricted to one or more of the states, and often they are further restricted within a state. Such a phenomenon is called *partial solubility*.

In addition to formation as solid solutions, alloys may display some chemical interaction between the component metals which brings about some new phase of intermetallic compound. The type of solid solution most commonly encountered in alloy work is the substitutional solid solution, in which the atoms of the component metals are either randomly mixed or occupy ordered atomic sites in the common lattice. In other solid solutions, the small atoms occupy interstices left between the closely packed larger atoms (see Section 1.7).

The phase transformation properties displayed by many different alloys are traditionally represented in the form of thermal equilibrium diagrams. (Some people prefer the name constitution diagrams, since equilibrium can only be reached by an infinitely slow rate of cooling which cannot be achieved in laboratory conditions.) Both binary and ternary equilibrium diagrams provide information on the existence of certain solutions, singly or in combination with others, with respect to temperature and composition. However, many equilibrium diagrams were produced long before X-ray techniques were applied, and the accuracy of these early diagrams must be viewed with a certain amount of caution.

The classical method of compiling a binary equilibrium diagram is to prepare a large number of alloys of the system at various compositions; each specimen is then heated to the required temperature, and this is followed by rapid quenching to preserve the phases present at elevated temperatures. Polishing, etching and subsequent microscopic examinations reveal the number and type of phases present at some specific temperature. In order to determine the position of phase boundaries, sophisticated cooling curve readings are also taken.

The techniques employed in classical equilibrium work are well developed in metallurgical laboratories, but a few criticisms could be mentioned here. For example, microscopic examination means that a series of visual inspections must be made to ascertain the presence or absence of certain phases. One obvious disadvantage is that a secondary phase, which is well dispersed in the lattice, may escape undetected. The Al–4% Cu shown on page 105 is such an example. The ageing in this alloy is quite well advanced, yet it is not visible under a microscope. Similarly, changes which occur in the solid solution, such as the order–disorder transformation of substitutional solutions, can also escape detection under the metallurgical microscope.

Perhaps the greatest limitation of microscopic investigations is that they provide no details about the possible structure of the phase (or phases) present. This can be overcome by the use of both classical and X-ray techniques side by side: each method then supplements and confirms the results obtained by the other. This principle was applied recently in investigations which used thermal methods to ascertain the solidus–liquidus lines, and X-ray techniques (confirmed micrographically) to fix the solid-phase boundaries accurately.

A useful technique for finding the phase boundaries is the disappearing-phase method. X-ray powder patterns of a series of alloys are made: the powder patterns of a single-phase alloy will show a single pattern; and for those alloys consisting of two phases, two patterns will be superimposed. It follows that, when the

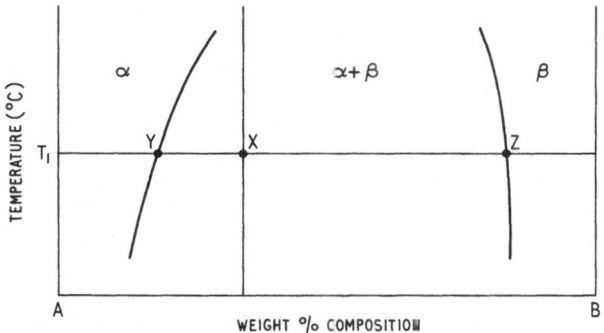

Fig. 7.1. The lever principle for the relative proportions of phases

boundary of the two-phase field is reached one of the patterns will gradually disappear. The limiting composition, at which a phase completely disappears, locates the boundary position. For accurate boundary determination, it is therefore necessary to prepare many specimens, with compositions crowded around the approximate boundary position. The number of samples to be examined can be reduced by the application of Gibbs' phase law. This law states that the relative amounts of phases co-existing in equilibrium at a given composition and temperature are inversely proportional to the distances to the corresponding phase boundaries. This rule is also known as the *lever principle.*

In Fig. 7.1, a portion of a hypothetical equilibrium diagram between metals A and B is shown. In a sample of composition X, both α and β phases exist and their weights W_α and W_β are related by the equation:

$$\frac{W_\alpha}{W_\beta} = \frac{\text{distance } XZ}{\text{distance } XY} \tag{7.1}$$

139

The relative amounts of phases present in a given sample may be deduced from intensity measurements of its powder pattern, since the intensity of each pattern is proportional to the quantity of the corresponding phase contained in the sample. In principle, it should then be possible to calculate the positions of the phase boundaries from the known composition and the proportions of phases. However, the difficulties encountered in this method are numerous. Firstly, its accuracy depends on the visual assessment of diffraction patterns to locate the disappearance or presence of a phase. Secondly, intensities do not always show linear proportionality to quantities present. Consequently, checking of results by independent means is always recommended.

An alternative technique, known as the parametric method, has been devised so that the intensity measurements, which often prove to be serious sources of error, can be avoided. The parametric method is based on Vegard's law, which states that the lattice parameters of substitutional solutions are continuous variables and are functions of composition only. The implication of this law is that a continuous variation of lattice spacings is to be expected whenever a change in the composition occurs within a single-phase system. However, if the composition of a sample lies outside the single-phase field, the two composing phases will be at their maximum solubility. Consequently, any change of composition in the

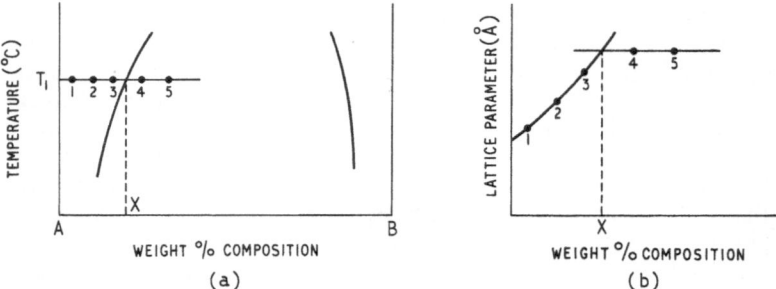

Fig. 7.2. *The parametric method of determining a solvus line*

sample, within the double-phase field, will not affect the composition of the individual phases, but will alter only the relative distribution of phases.

Fig. 7.2 shows diagrammatically how the parametric method is applied. A series of alloy samples, numbered one to five, are prepared and are heated together to a temperature T_1. After a sufficiently rapid quenching, a powder is produced by careful filing. To eliminate the residual cold work caused by the filing, the powder sample is subjected to a low-temperature recovery anneal. Following this treatment, a set of powder photographs of the samples is prepared. The lattice parameters are then calculated and plotted, as illustrated in Fig. 7.2(b); the critical composition X of the solubility limit at the temperature T_1 is given by the projection of the intersection of the two portions of the graphs on to the horizontal axis.

The complete derivation of the solvus line can be simplified if the composition of the sample is chosen within the two-phase field, and if the specimen is heated to achieve perfect homogenisation. After extensive heating, the block specimen is rapidly quenched, and the powder sample is prepared and submitted to X-ray investigation in the usual manner. The resulting diffraction powder pattern will then contain the information needed to find the lattice parameters of both component phases. The experiment is repeated with the sample heated to other, subsequently lower, temperatures. If the terminal solid solution is such that its saturation composition is temperature-dependent (which nearly always happens), the lattice parameters will vary slightly. A calibration curve may be made by means of various single-phase compositions for which the lattice parameters have been calculated. The combination of the two sets of data will then provide the variation of composition with temperature.

The need for care in the preparation of block samples and subsequent powders cannot be sufficiently emphasised. It is known that even traces of impurities present in a sample may falsify results considerably. When the block samples are heated, different lengths of time at a constant temperature must be used so that the correct period (after which no appreciable change occurs) can be found. The preparation of powders is an art in itself. The powders should be prepared either in vacuo or in an inert gas to avoid oxidation and atmospheric contaminations. In certain alloys, silicon from dust contamination may suppress second-phase precipitation to a considerable extent.

Reliable thermal equilibrium diagrams are, of course, of paramount industrial importance for choosing alloys and for practical heat treatment selection. On the other hand, it often seems that excessive accuracy only demonstrates high-class laboratory facilities, without necessarily improving industrial techniques.

For instance, most engineering alloys used in industry are contaminated beyond recognition; in addition, if the phase boundary is displaced by 0·2% in either direction in a given equilibrium diagram, the quality of the end product will hardly be affected.

Both the disappearing-phase technique and the parametric method are equally applicable to the determination of ternary systems. The almost universally accepted convention now is to plot the phases of a three-component system in an equilateral triangle, as shown in Fig. 7.3. The complete phase equilibrium can only be shown in a three-dimensional diagram, because there are four variables to consider. In practice, however, the equilateral triangles serve as isothermal sections, and a few such sections fortunately are sufficient to provide the necessary information. The standard way of plotting ternary equilibrium diagrams is to represent the pure metals A, B and C by the three corners of the triangle. The binary alloys are then represented by points lying on the perimeter of the triangle. The composition at a point P is given by the relationships:

$$\left. \begin{array}{l} \dfrac{\text{amount of } A}{\text{amount of } B} = \dfrac{PX}{PY} \\[2ex] \dfrac{\text{amount of } A}{\text{amount of } C} = \dfrac{PX}{PZ} \end{array} \right\} \qquad (7.2)$$

The lines PX, PY and PZ are drawn perpendicular to the sides BC, CA and AB respectively, as shown in Fig. 7.3; their sum is constant for all positions of P and is taken as 100%.

In ternary diagrams, single-phase, double-phase and triple-phase fields can exist together. Fig. 7.4 is an isothermal section of a hypothetical ternary diagram. The relative amounts of phases present in the double-phase field and in the triple-phase field can be calculated by means of a modified lever principle. In double-phase fields, 'tie' lines can be drawn; along these lines, the relative amounts of phases change but their compositions remain constant. The tie lines can be drawn as a series of straight lines that never cross each other, but fan out within the triangle.

The line EDF in Fig. 7.4 is a tie line. Thus the composition of alloy D is given by the relationship:

$$\frac{\text{amount of } \alpha}{\text{amount of } \beta} = \frac{DF}{DE} \qquad (7.3)$$

When composition Q (Fig. 7.4) lies within the three-phase field, a

142

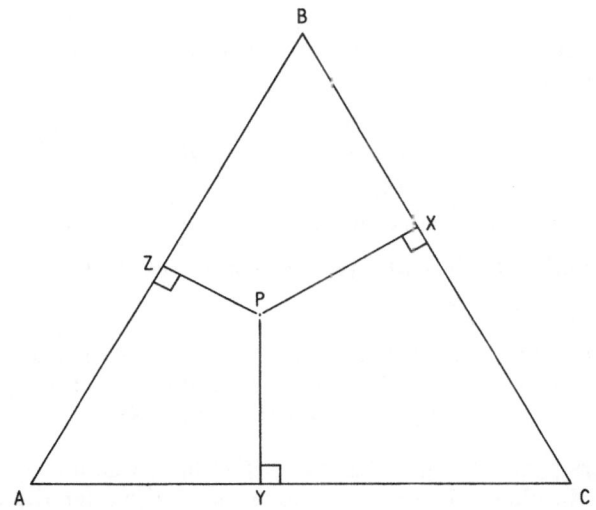

Fig. 7.3. Equilateral triangle of a ternary system

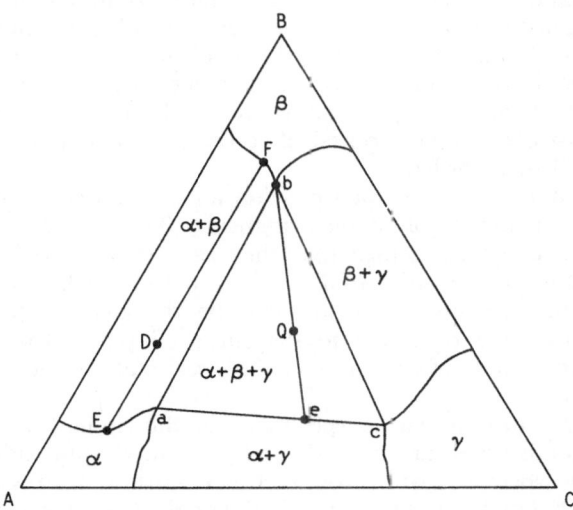

Fig. 7.4. Isothermal section of a hypothetical ternary diagram

line is drawn through Q and one corner of the field (say b) to cut the opposite side of the field at e. The proportions are then given by the equations

$$\left.\begin{array}{l} \dfrac{\text{amount of } \beta}{\text{amount of } \alpha + \gamma} = \dfrac{Qe}{Qb} \\[2mm] \dfrac{\text{amount of } \alpha}{\text{amount of } \gamma} = \dfrac{ec}{ea} \end{array}\right\} \qquad (7.4)$$

where a and c are the other two vertices of the field.

7.3 MEASUREMENT OF STRESSES IN POLYCRYSTALLINE METALS BY X-RAY DIFFRACTION

In general, stresses are classified as either residual or applied stresses. Residual stresses are those which exist after the applied stress is removed; they can be induced in a variety of ways, and are commonly caused by manufacturing processes, such as welding or casting.

One of the most common methods of determining the deformation (and thus the stress) in a material is to attach an electrical or mechanical strain gauge and then to observe the minute dimensional changes caused by an applied stress. However, this technique is quite impracticable in some circumstances: if the material already contains some residual stresses, only those residual stresses superimposed after the attachment of the strain gauge will be recorded.

X-ray diffraction methods provide an alternative, in that they are suitable for absolute measurements. These diffraction techniques have the advantage that they can be used to investigate small areas (of the order of 1 mm^2), and to scan larger areas so that inhomogeneities of strain can be detected. Although the observations are restricted to the surface layer of the material, the surface results often represent the stress state of the rest of the specimen.

Consider now a tensile specimen, as shown in Fig. 7.5. The applied loads are so arranged that they produce a pure tensile load only. The specimen with cross-sectional area A is stressed in its longitudinal direction by a force P; the applied stress σ_l is therefore equal to P/A. This stress σ_l produces an amount of strain ε_l in the longitudinal direction only. By definition, the strain is given by the relationship

$$\varepsilon_l = \frac{L_e - L_0}{L_0} \qquad (7.5)$$

where L_0 and L_e are the original and extended lengths respectively.

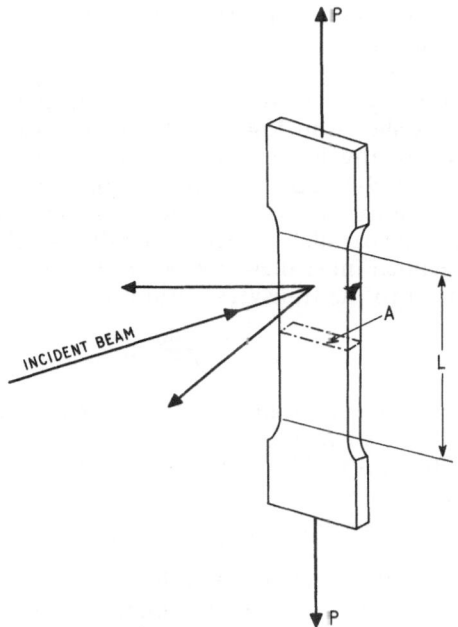

Fig. 7.5. A specimen in pure tension

If the stress is chosen within the limit of proportionality, then the applied stress is related to the strain by the equation

$$\sigma_l = E\varepsilon_l \qquad (7.6)$$

where E is Young's modulus.

The concept of constant volume prescribes that a reduction in the cross-sectional area A will accompany any extension in L. The transverse strain ε_t is therefore given by the expression

$$\varepsilon_t = \frac{A_e - A_0}{A_0} = -\nu\varepsilon_l \qquad (7.7)$$

145

where A_0 and A_e are the original and contracted areas. The specimen is assumed to behave isotropically—i.e. the longitudinal and transverse strains are related by Poisson's ratio v of the material.

X-ray diffraction techniques can be used to measure the changes which occur in the overall length of the specimen within the elastic range: the techniques are quite suitable for detecting the minute changes which occur in the spacings of the atomic planes as a result of applied stresses. If the longitudinal strain ε_l is to be evaluated, the slight increase in the spacing of the atomic planes perpendicular to the longitudinal axis of the specimen must be measured. An X-ray beam directed perpendicular to the axis of the specimen, as shown in Fig. 7.5, cannot record the changes in ε_l; this is because the incident beam and the lattice planes are nearly (or completely) coplanar, and the diffracted beam is thus directed along the axis of the specimen. Back-reflection measurements from planes parallel, or nearly parallel, to the longitudinal axis of the specimen must therefore be used. The change in lattice spacing is related to the transverse strain by the equation

$$\varepsilon_t = \frac{d_e - d_0}{d_0} \qquad (7.8)$$

where d_0 and d_e are the original and extended spacings of planes reflecting at normal incidence.

Equations 7.6, 7.7 and 7.8 can be combined to give

$$\sigma_l = -\frac{E}{v}\left(\frac{d_e - d_0}{d_0}\right) \qquad (7.9)$$

which is the required relationship between the experimentally determined lattice spacings and the unknown longitudinal stress. It must be remembered that Equation 7.9 is only valid if the applied stress is within the limit of proportionality; otherwise, the linear proportionality implied by E/v is no longer acceptable.

The unstrained value of d_0 in Equation 7.9 should be obtained on a stress-free portion of the specimen – and care should be taken in deciding which part of the specimen is stress free. If no other evidence is available, a small sample of powder could be prepared and subjected to extensive annealing treatment. The strain-free spacings could then be determined by the methods of indexing and parameter calculation described in Chapter 6.

The elastic constants used in the stress determination should correspond to the indices of the reflecting plane. It is well worth remembering that Young's modulus is a fourth-rank tensor pro-

146

perty. For single crystals of iron, Goens and Schmid found different values of Young's modulus to be:

$$E_{100} = 13,500 \text{ kg/mm}^2$$

$$E_{111} = 29,000 \text{ kg/mm}^2$$

The ratio E_{111}/E_{100} for copper is about 2·5, and for aluminium the same ratio drops to between about 1·4 and 1·2. In view of this large anisotropy, the question arises as to which value of E should be used in X-ray stress determinations. One useful procedure is to compile a calibration curve by loading a standard specimen from the same batch of material as the one under investigation. Such standard specimens are normally made as flat beams or circular split rings.

Since the magnitude of the applied load and the geometry of the calibration piece are known, the stresses can be readily calculated from conventional elastic theory. If the measured values of d_e and d_0 and the calculated value of σ_l are then substituted into Equation 7.9, the ratio E/v (the stress factor) may be calculated.

For biaxial stresses, the system becomes more complicated by an extra dimension. According to elastic theory, in a stressed body there are three mutually perpendicular directions each of which is perpendicular to a plane that is free of shear stresses. These mutually perpendicular directions are called the principal directions; the stresses acting along them are called the principal stresses, and are denoted by σ_x, σ_y and σ_z. If the stress σ_z is taken as zero on the free surface of a body, the strain ε_z is given by the equation

$$\varepsilon_z = -\frac{v}{E}(\sigma_x + \sigma_y) \tag{7.10}$$

The value of ε_z can be determined from the diffraction pattern which is produced by an incident X-ray beam normal to the free surface of the body. In general, a diffraction pattern can provide information about the sum of the principal stresses in a biaxial stress system.

Since the methods of determining biaxial and uniaxial stresses both rely on calculations of interplanar spacings, the experimental techniques used for obtaining diffraction patterns are extremely important. In stress determinations, back-reflection cameras are invariably used — primarily because the absorption of X-rays by large cast or fabricated specimens is usually too high for X-ray transmission photographs to be taken. In addition, the minute

147

parametral changes caused by stresses shift the high-angle lines (which are used in the back-reflection technique) more than the low-angle ones (which are obtained with transmission photographs). The geometry of each technique is therefore arranged so that the incident beam is perpendicular to the axis of the applied load for a uniaxial stress, and perpendicular to the principal plane for a biaxial stress.

If the crystals under investigation tend to be large, as often happens in aluminium and magnesium alloys, the diffraction rings will be spotted and this will adversely affect the accuracy of the stress determinations.

Because of the geometry of back-reflection and also because the amount of strain varies with the indices of reflection, interplanar spacing variations are determined from one diffraction ring only. Consequently, since spacing variations cannot be averaged as in lattice parameter calculations, other precautions are necessary for maximum reliability of results. For instance, it is very important to know, and to be able to reproduce accurately, the film-to-specimen distance, since this affects the diameter of the diffraction ring. Two basic methods of determining this distance are currently in use. In one, a thin layer of strain-free powder, usually either gold or aluminium paint, is deposited on the surface of the specimen, and their diffraction patterns are recorded simultaneously. From the measured diameter of a diffraction ring due to the strain-free powder and from the known lattice parameter, the film-to-specimen distance can be calculated. The alternative method is to provide a ground distance-piece which fits closely between the specimen and the recording film, and enables the operator to place the specimen accurately in a precisely defined position.

Apart from the normal incidence methods of measuring stresses, numerous other techniques are presented and discussed in the original literature. For the successful application of all of these methods, the stresses and strains must be small in magnitude and should be within the elastic limit.

7.4 MEASUREMENT OF GRAIN SIZE IN POLYCRYSTALLINE METALS

Perhaps the most important single factor affecting the mechanical and metallurgical properties of worked or cast polycrystalline metals is the average grain size of the polycrystalline mass. As a

148

rule, a reduction in the average grain size produces a corresponding increase in the mechanical strength. For instance, the yield stress of 70/30 brass is about 3·3 tons/in^2 at an average grain diameter of 0·2 mm, increasing to about 9 tons/in^2 when the average grain diameter is about 0·001 mm. In pure iron, a reduction in grain size results in a considerable increase in hardness: a grain size of 1·4 mm corresponds to about 70 BHN, and a reduction in grain size to 0·23 mm results in an increase in hardness to 100 BHN. According to Bragg's experimental deductions, the yield stress of a metal varies with the average grain size as k/s or $k/(s-s_0)$ where k is a constant of proportionality, and s and s_0 are the mean and the limiting crystallite size.

It can be seen that the measurement of average grain size is important if certain mechanical properties of engineering products are to be predicted. For practical purposes, grain sizes can be measured in one of two systems: in one system, the mean grain diameter is given; and in the other, the number of grains per unit area is given. The second system is currently preferred, because it is easier to count grains than to measure them. There are certain anomalies which arise in the measurement of grain sizes. The most obvious limitation is that an average grain size is independent of the shapes of the grains — and these may vary considerably. In other words, the variation of strength with size can only be predicted when the grains are nearly equiaxial and of almost uniform size.

In the classical method of measuring grain size, a polished and etched specimen undergoes a fairly simple optical examination. The number of grains in a known area of the microscopic field is counted, and the average grain size is thence calculated in terms of unit area. To avoid the actual counting of grains, the American Society for Testing Materials has prepared a visual grain size index card which, by direct comparison, gives the average number of grains per square inch.

In many industrial problems, polishing and etching operations are considered as destructive testing because of the damage they produce on commercial surfaces; the surface finishes are adversely affected, and consequently optical methods of grain counting cannot be employed. For practical work of this nature, X-ray methods of determining grain sizes are frequently the most useful and economic. The simplest (and often the most suitable) method depends on the visual comparison of a set of calibrated photographs and the photographs obtained from the specimen under investigation.

149

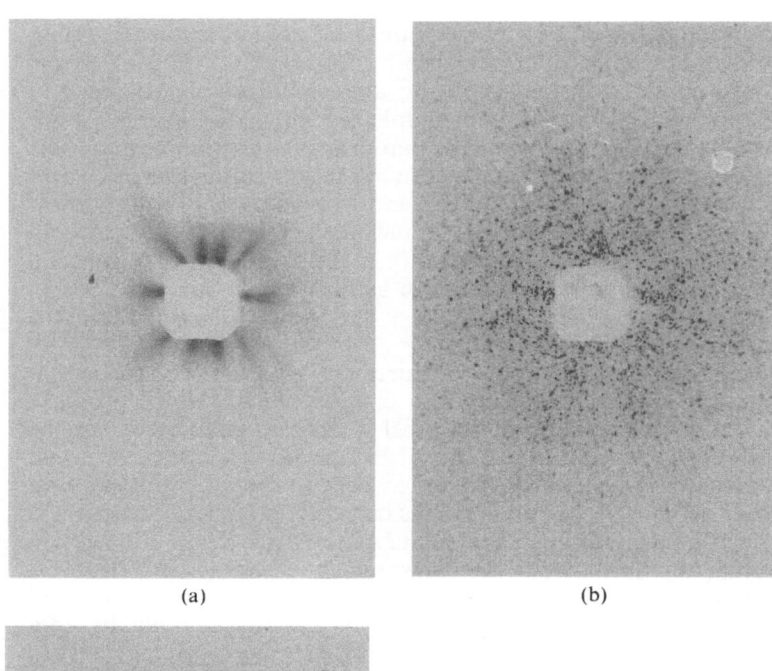

(a)

(b)

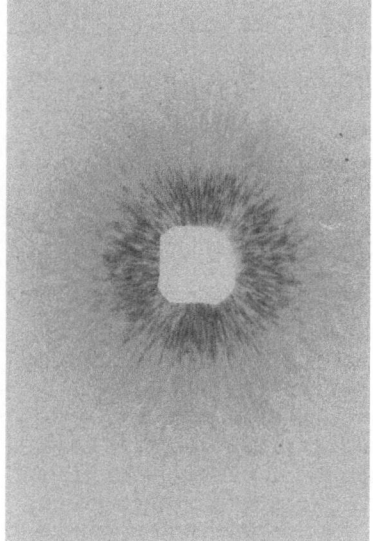

(c)

Fig. 7.6. Transmission patterns of a commercially pure rolled aluminium strip subjected to various annealing treatments: (a) half hard as rolled; (b) recrystallised at 750°C in a travelling furnace at a rate of 1·5 cm/min; (c) recrystallised at 660°C in a travelling furnace at a rate of 3 cm/h

When an X-ray beam falls upon a single crystal, a defined number of diffraction spots will occur. As the number of crystals is increased, the incident X-ray beam suffers multiple diffraction with a resulting superimposition of many Laue patterns (as in Fig. 7.6). When the average crystal size becomes about 10^{-4} cm, the spots merge into continuous Debye–Scherrer rings. Further reduction, to between 10^{-4} cm and 10^{-5} cm, produces no visible change in the diffraction pattern; but below a grain diameter of 10^{-5} cm, a marked broadening of Debye–Scherrer rings occurs.

An individual diffraction spot may provide valuable information about grain size, since the shape and size of a diffraction spot is related to the grain size of the reflecting crystal. For grain diameters between 10^{-4} cm and 10^{-5} cm, the relationship between the vertical spot size and the measured grain diameter is linear. As a result, calibration curves can be constructed for given cameras with the aid of samples of known grain sizes near the limiting values; diffraction spot measurements can be plotted against average grain diameters, and a straight line can then be drawn.

When the grain size falls below the 10^{-5} cm limit, measurements are based on line broadening. In general, the width of Debye–Scherrer rings depends on the intrinsic broadening β (in radians), on a measure b of the divergence of the incident X-ray beam, and on some geometrical factors of the camera. The relationship between β and the average crystallite size s is found to be

$$\beta = \frac{\lambda}{s \cos \theta} \qquad (7.11)$$

where λ is the wavelength of the incident radiation.

The problem is to evaluate β from the measured widths B of the diffraction rings. Both theoretical and experimental observations lead to the relationship:

$$\beta = [(B-b)(B^2-b^2)^{\frac{1}{2}}]^{\frac{1}{2}} \qquad (7.12)$$

The divergence b of the incident beam may be found experimentally from Debye–Scherrer rings obtained from powders of known particle size: Equation 7.11 can then be solved for β and Equation 7.12 for b.

It must be remembered that these methods of grain size determination are applicable only when the specimen is stress free. This is particularly important when line broadening is considered, since plastic strain causes considerable line broadening. Until recently, one method of producing fine powders of ionic salts well below the 10^{-4} cm level was to ball mill them for some length of time; it was generally assumed that ionic salts were not plastically deformable in ball mills. However, comparison of line broadening measurements and electron microscope results has proved that pulverised ionic salts do readily deform plastically in ball mills, and consequently the validity of earlier experimental conclusions reached from line broadening measurements is in serious doubt.

8

The texture of polycrystalline wires and sheets

8.1 THE TEXTURE OF POLYCRYSTALLINE AGGREGATES

Metals, minerals and other polycrystalline substances which have undergone severe deformation in engineering processes display varying amounts of preferred orientation. The term 'preferred orientation of texture' is used to describe a certain alignment of crystallographic planes and directions in a preferred manner with respect to the direction of maximum deformation. Preferred orientation is also present in solidification processes, particularly when dendritic growth conditions are maintained. When the preferred orientation is a result of deformation, it is very dependent on the slip and twinning systems in the metal; but when orientation results from solidification, it depends mainly on the crystal structure.

Detection of preferred orientation prior to processing is of prime interest in industrial work. The anisotropy of some physical properties, which is occasionally sufficient to cause failure of various manufacturing processes, can usually be traced back to the formation of preferred orientation. The mechanical properties of individual crystals of metals are known to be anisotropic, but in random aggregates the directional variations average out and the aggregates tend to display some degree of isotropy.

The nature of textures developed during deformation may change when the metal is subjected to annealing. In industrial processing, such as hot or cold rolling, extrusion or forging, it is not easy to distinguish between the textures produced at various stages of temperature. If the whole of the process is carried out above the recrystallisation temperature, the resulting preferred orientations are much less definite than those usually expected from severe cold work.

One of the simplest orientation textures is found in drawn or extruded wires. Other fibrous materials, such as synthetic fibres, cotton, wool and silk, display a texture similar to that of drawn or extruded metallic wires. In order to describe such simple textures accurately, it is sufficient to define a crystallographic axis common to all grains and parallel to the fibre axis; all other axes are then randomly orientated around the fibre axis. Fig. 8.1 shows a

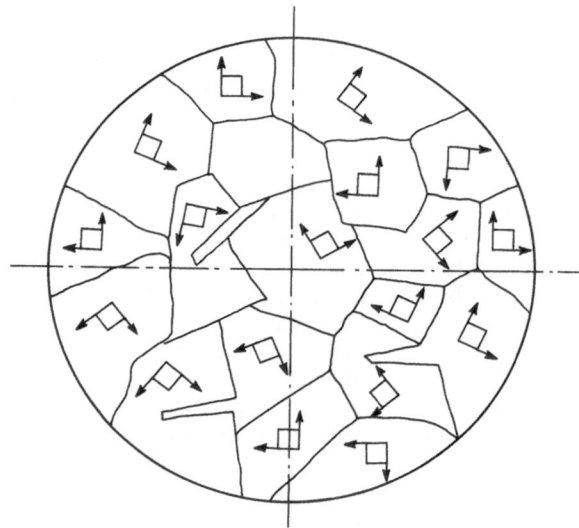

Fig. 8.1. Ideal orientation of a drawn or extruded wire

diagrammatic cross-section of a wire, looking along the wire axis. The small cubes indicate the orientation of the grains making up the wire. In the idealised texture, one edge of each cube (the $\langle 100 \rangle$ direction) is parallel to the wire axis, while the other two edge directions lie in the plane of section (azimuthal orientation) and are randomly orientated with respect to a transverse axis. Textures of this description are called *simplex fibre textures*.

Sometimes, there are two common crystallographic axes orienta-

154

ting themselves parallel to the axis of the wire; textures of this nature are called *duplex* textures.

In duplex textures, it is usual to give the percentage of grains aligning themselves in a preferential manner (on the basis that ideally 100% of the grains would align themselves in this way). Table 8.1 shows the percentage distribution of grains in some face centred cubic metals of duplex fibre texture.

Table 8.1.

Some Duplex Wire Textures

Metal	Percentage grains	
	$\langle 100 \rangle$ direction parallel to wire axis	$\langle 111 \rangle$ direction parallel to wire axis
Copper	40	60
Gold	50	50
Silver	75	25

The texture of rolled sheet could again be described in terms of deviations from an ideal orientation. The ideal orientation is normally defined by the (hkl) plane lying parallel to the rolling plane, and the $[uvw]$ direction lying parallel to the rolling direction. For example, if copper is given a straight, non-reversal rolling treatment, its ideal orientation is a $(110)/[112]$ texture. For body centred cubic α-iron, the resulting ideal orientation after rolling is $(100)/[011]$. In rolled textures, it is not unusual to find duplex or triplex textures. The ideal texture of rolled brass, for example, is $(100)/[011]$; $(112)/[110]$; $(111)/[112]$. It is conceivable that, in real situations, not all grains will occupy their prescribed orientation positions as accurately as the ideal orientation would suggest. To indicate the amount of 'throw', the angular deviations are sometimes given.

It has long been recognised that the description of orientation textures, especially of sheets and other rolled products, in terms of ideal orientation plus angular deviation is very unsatisfactory. As an alternative, Wever devised the *pole figure*, which provides a complete description of orientation texture. The Wever pole figure is very closely associated with the stereographic projection of face normals, and consists of areas of various pole densities. It is drawn to show the distribution of a chosen face normal—usually either the $\{111\}$ or the $\{100\}$ type poles. The construction

155

of a pole figure will be described in Section 8.3 in connection with the textural investigations of rolled sheets.

The determination of preferred orientation is greatly facilitated by the use of X-rays. Because of the rapidity and relative simplicity of X-ray techniques, their industrial application is much wider than that of optical methods based on etch-pit orientation principles. The next two sections show how the orientation texture of wires and of sheets can be found.

8.2 DETERMINATION OF THE FIBRE AXIS OF A WIRE

The simplest experimental arrangement makes use of a beam of monochromatic radiation perpendicular to the axis of a vertical wire. Some mechanical means of rotating the wire around its axis must also be provided. Diffraction patterns can be recorded on either flat or cylindrical films: it is simpler to interpret the

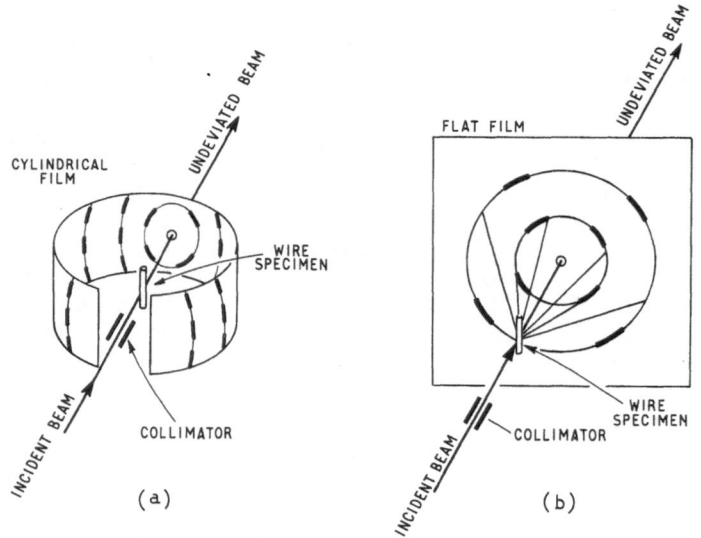

Fig. 8.2. Experimental arrangement for recording diffraction patterns of a wire: (a) cylindrical film recording; (b) flat film recording

diffraction pattern recorded on a flat film, but a cylindrical film provides a greater recording range. The experimental arrangement is shown diagrammatically in Fig. 8.2.

156

Consider first the diffraction pattern which would be produced by an ideal wire; if all the grains in the wire were perfectly aligned with a crystallographic axis parallel to the wire axis, the diffraction pattern would also be idealised. Fig. 8.3 shows an incident mono-chromatic beam impinging on a plane (hkl) at point O. The face

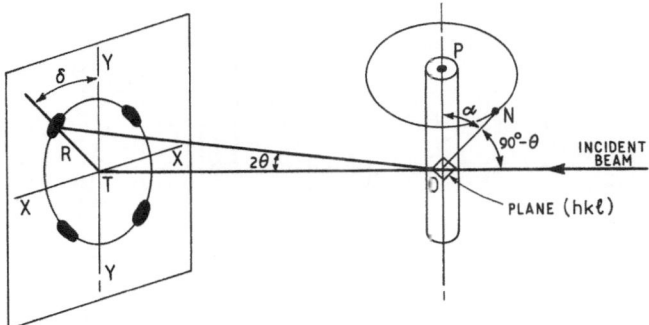

Fig. 8.3. Formation of an idealised diffraction spot on a flat film

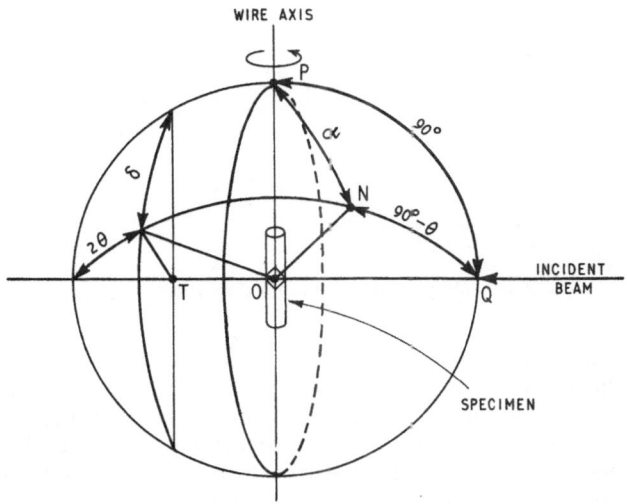

Fig. 8.4. The spherical relationship between α, θ and δ of Fig. 8.3

157

normal ON of the plane (hkl) makes an angle α with the axis OP of the wire. If the wire is rotated around the axis OP, the face normal ON traces the surface of a cone with half-angle α. In a complete rotation of 360°, the (hkl) plane moves four times into a position which satisfies Bragg's reflection condition; thus four spots on the Debye–Scherrer ring will be formed. The angle δ at

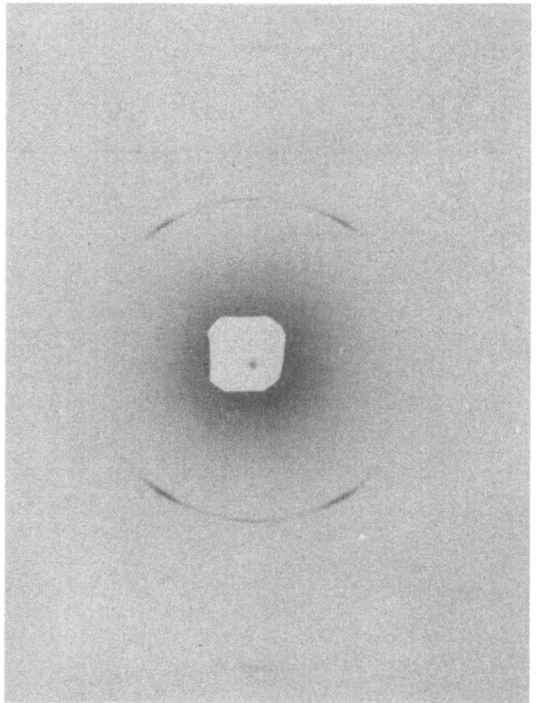

Fig. 8.5. *Diffraction pattern of a drawn tungsten wire*

which the reflection spot occurs is a function of both α and θ, while the radius R is a function of θ and the distance OT. If the incident beam is perpendicular to the wire axis, the reflection spots will appear to be symmetrical with regard to the X and Y axes.

In general, three special conditions arise which modify the four points of the idealised diffraction pattern. When $\alpha = 90°$, the diffracting plane is perpendicular to the incident beam and consequently only two spots will appear in the pattern, lying at the intersection of the X axis and the Debye–Scherrer ring. When $\alpha = 0°$, the diffracting plane is parallel to the incident beam and no reflection can occur. Finally, when $\alpha = \theta$, only two reflections can again occur and these will appear at the intersection of the Y axis and the Debye–Scherrer ring.

The relationship between the variables α, θ and δ may be deduced by the application of spherical trigonometry. In Fig. 8.4, a reference sphere is drawn around point O and the geometry of Fig. 8.3 is reproduced within the sphere. The position of point N is chosen so that it satisfies the Bragg condition. From the spherical triangle PNQ, the cosine relationship yields the equation

$$\cos \alpha = \cos(90° - \theta) \cos 90° + \sin(90° - \theta) \sin 90° \cos \delta \quad (8.1)$$

which can be further simplified to

$$\cos \alpha = \cos \theta \cos \delta \quad (8.2)$$

Fig. 8.5 shows an X-ray diffraction photograph of a tungsten wire of 0·004 in diameter. Copper $K\alpha$ radiation was employed at 35 kV and 13 mA, with an exposure time of 90 min. A flat film was used, and the film-to-specimen distance was fixed at 2·5 cm. Tungsten is a body centred cubic element with lattice parameter $a = 3·1648$ Å. The most noticeable feature of the photograph is that the Debye–Scherrer rings are quite clear, with very dark areas superimposed on them. The pattern is evidently in transition between the uniformly intense rings of an ideally random powder photograph and the localised darkened spots of an ideal texture photograph. This phenomenon is predictable, since some of the grains will not occupy ideally orientated positions but will remain randomly orientated.

The azimuthal angle δ (see Fig. 8.3) is measured between the centre of the most intense segment of a ring in the photograph and a vertical fiducial line drawn on the film. The first step in finding the crystallographic direction which lies parallel to the wire axis is to index the partial Debye–Scherrer rings. From Fig. 8.5, the radius R_1 of the inner ring is found to be 2·242 cm, and the radius R_2 of the outer ring is 4·237 cm. The Bragg angle θ can be calculated from the equation

$$\tan 2\theta = \frac{R}{D} \quad (8.3)$$

where R is the radius of the diffraction ring and D is the distance

159

between the specimen and the film. If the measured radii are substituted into Equation 8.3, then

$$\tan 2\theta_1 = \frac{2 \cdot 242}{2 \cdot 5} = 0 \cdot 8969$$

and

$$\tan 2\theta_2 = \frac{4 \cdot 237}{2 \cdot 5} = 1 \cdot 6945$$

Therefore

$$\sin^2\theta_1 = 0 \cdot 1276$$

and

$$\sin^2\theta_2 = 0 \cdot 2460$$

The lattice spacing d for a cubic crystal is given by the equation:

$$d = \frac{a}{(h^2 + k^2 + l^2)^{\frac{1}{2}}} = \frac{a}{\sqrt{N}} \tag{8.4}$$

Equating d of Equation 8.4 to the lattice spacing obtained from the Bragg equation gives

$$\frac{\lambda}{2 \sin \theta} = \frac{a}{\sqrt{N}} \tag{8.5}$$

Solving for N

$$N = \left(\frac{2a}{\lambda}\right)^2 \sin^2\theta \tag{8.6}$$

Therefore, by substitution of a, λ and $\sin^2\theta_1$ and $\sin^2\theta_2$,

$$N_1 = \left(\frac{2 \times 3 \cdot 1648}{1 \cdot 5418}\right)^2 \times 0 \cdot 1276 \simeq 2$$

and

$$N_2 = \left(\frac{2 \times 3 \cdot 1648}{1 \cdot 5418}\right)^2 \times 0 \cdot 2460 \simeq 4$$

The indices of the inner ring are therefore 110, and those of the outer ring are 200. The azimuthal angle δ was measured from the photograph with the aid of a protractor. The data for the two rings are summarised in Table 8.2.

Angle α is now calculated by substitution of θ and δ into Equation 8.2:

$$\cos \alpha_{110} = \cos 20\cdot6° \cos 57° = 0\cdot5010$$

or

$$\cos \alpha_{110} = \cos 20\cdot6° \cos 90° = 0$$

Therefore α_{110} is equal to 60° or 90°. Similarly, for α_{200},

$$\cos \alpha_{200} = \cos 29\cdot5° \cos 90° = 0$$

Hence α_{200} is equal to 90°.

Table 8.2.
Data for the Determination of the
Fibre Axis of a Tungsten Wire

Ring	θ	N	Indices of reflection	δ
Inner	20·6°	2	110	57°, 90°
Outer	29·5°	4	200	90°

From Fig. 8.4, the angle α is the angle subtended between a face normal ON and the axis of the wire. Therefore the direction of the wire axis is the direction which lies at either 60° or 90° to the [110] direction, and at the same time lies at 90° to the [200] direction. From consultation of tables of interplanar angles in the cubic system (see Table 5.1), it can be found that the only possible direction indices are the ⟨110⟩ set. It can therefore be concluded that the tungsten wire used in the experiment has a pronounced texture, in which the grains orientate themselves preferentially and exclusively in the ⟨110⟩ direction.

Fig. 8.6 shows a set of photographs of the diffraction pattern of drawn copper wire (FCC, $a = 3\cdot6153$ Å). The experimental conditions were similar to those used for the tungsten wire, except that a cylindrical camera of 3 cm radius was employed. The photograph shown in Fig. 8.6(a) was taken when the copper wire had received no surface treatment whatsoever; the other two diffraction photographs were taken after etching in a solution of ferric chlorite in alcohol had reduced the diameter of the same specimen in steps of 0·005 in. The most noticeable feature of the photographs is a marked sharpening of texture towards the centre

161

of the wire. As this phenomenon occurs equally in drawn, extruded and swaged wires, it is most probably due to the differential radial strain distribution imposed on the wire.

The interpretation of a cylindrical film is entirely analogous to the method described above for a flat film example. However, only the horizontal diameters of the diffraction rings should be measured because the other diameters are distorted by the cylindrical geometry. Transparent charts with constant θ and constant δ lines can be used to simplify the work of interpreting the photographs. A suitable chart prepared for the 3 cm radius universal camera is available from The Institute of Physics and The Physical Society, London.

The most pronounced texture, shown in the photograph of Fig. 8.6(c), will now be analysed. A constant θ–δ chart was superimposed on the photograph, and the data obtained for the first four lines (working from the centre outwards) are shown in Table 8.3. A closer examination of the diffraction rings of Fig. 8.6 reveals more than the expected four points: instead, eight concentrations of intensity can be seen in the innermost ring, reducing to six in the next two rings. These high-intensity points are not easily visible with the unaided eye, because the angular spread of the spots is so large that they appear to overlap each other. This duplex pattern can be explained by the theory that two sets of grains

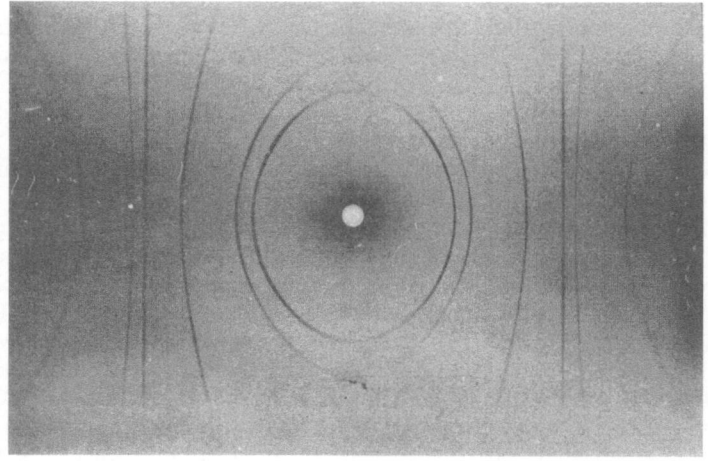

(a)

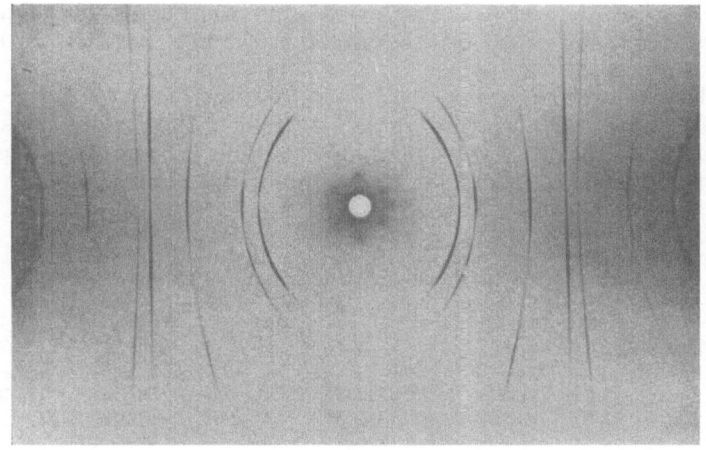

(b)

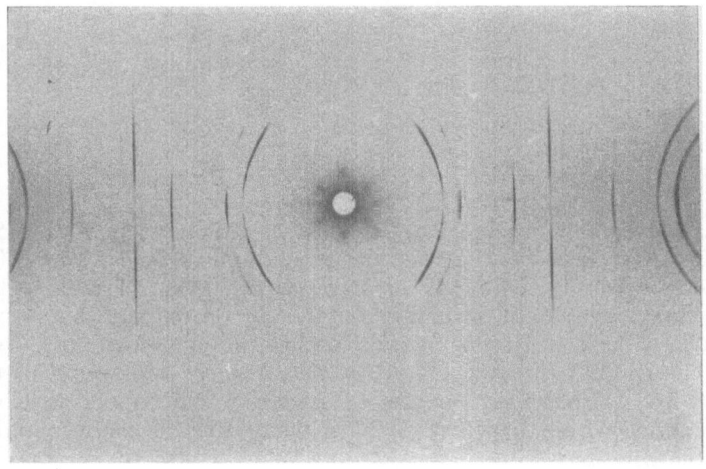

(c)

Fig. 8.6. Diffraction patterns of drawn tough pitch copper wire: (a) 0·015 in diameter; (b) 0·010 in diameter; (c) 0·005 in diameter

are probably present in the wire, each set having a different crystallographic axis orientated more or less parallel to the direction of the wire axis.

The values of α_1 and α_2 calculated from the measured values of θ, δ_1 and δ_2 are shown in Table 8.4. If all of the α values closely match two specific crystallographic directions of the cubic system, then the assumption of duplex texture can be taken as correct.

Table 8.3.

Data for the Determination of the Fibre Axis of a Copper Wire

Line	θ	Indices of reflection	cos θ	δ_1	δ_2	cos δ_1	cos δ_2
1	22·5°	111	0·9239	53°	68°	0·6018	0·3746
2	25·1°	200	0·9048	53°	90°	0·6018	0
3	37°	220	0·7986	32°	90°	0·8480	0
4	45°	311	0·7071	73°	—	0·2924	—

Table 8.4.

Determination of the α Values of a Copper Wire

Line	cos α_1	cos α_2	α_1	α_2
1	0·5550	0·3455	56°	70°
2	0·5450	0	57°	90°
3	0·6750	0	47·5°	90°
4	0·2065	—	78°	—

The table of interplanar angles for the cubic system (Table 5.1) shows that the [111] direction lies at an angle of 54·7° to the [100] direction and at an angle of 70·5° to the [111] direction. The tabulated α values of the first line of the diffraction pattern approximate to these theoretical angles, within the experimental margin of error. Thus the conclusion reached from consideration of the first line is that the two sets of grains of the copper wire are orientated with either their $\langle 111 \rangle$ or their $\langle 100 \rangle$ directions parallel to the wire axis.

The same table of interplanar angles shows that the [220] direction lies at 90° to the [111] direction and at 45° to the [100] direction. The α values calculated for the third line again correspond

164

to these angles, within the limits of experimental error — confirming the results obtained from the first line. The fourth line [the diffraction ring due to the (311) atomic planes] provides only one α value of 78°. The table of interplanar angles shows that the angle between the (311) plane and the (100) plane is 72·5°, and the angle between the (311) plane and the (111) plane is 80°. It can be assumed that the single α value of 78° is probably in error on account of the considerable spread of two closely overlapping segments. Since the third and fourth lines thus confirm the results obtained from the first line, it can safely be concluded that the duplex texture is such that the grains orientate preferentially with either their $\langle 111 \rangle$ or $\langle 100 \rangle$ directions parallel to the wire axis.

The percentage of the grains which belongs to each set can only be decided after careful intensity studies of the diffraction photograph. The governing principle is that the greater the number of grains, the more intense the corresponding spot will be. A visual examination of the diffraction rings for copper reveals some of the intensity differences, but is not sufficient to indicate the known 60/40 distribution of grains.

8.3 THE TEXTURE DETERMINATION OF ROLLED SHEETS

The principles of texture determination discussed in Section 8.2 are equally applicable to steels and other rolled products. However, it must be taken into consideration that the texture is not formed by a radially symmetrical operation, such as extrusion or wire drawing, but is the result of a biaxial deformation system applied to the sheet in a linearly symmetrical manner. The grains of rolled sheets therefore occupy an orientated texture, in which one axis preferentially points in the rolling direction while an atomic plane preferentially lies parallel to the rolling plane. The ideal orientation of sheets is hence described in terms of the (hkl) plane and the $[uvw]$ direction which lie respectively in the plane and direction of rolling. Ideal orientations are nearly always found by plotting and interpreting the pole figure. For cubic metals, it is customary to present the $\{100\}$, $\{110\}$ and $\{111\}$ type pole figures separately from the less descriptive ideal orientations.

The simplest method of plotting pole figures is to prepare three photographs with the specimen sheet arranged in the following three different ways.

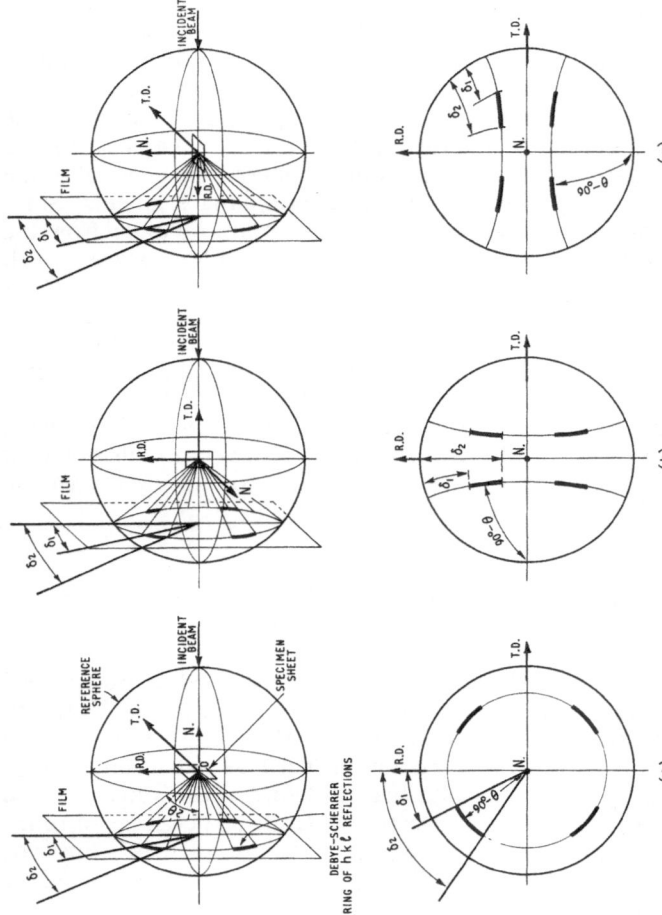

Fig. 8.7. Location of specimens relative to the incident X-ray beam, and the corresponding stereographic projections of the Debye–Scherrer rings: (a) N. parallel to incident beam, R.D. vertical; (b) T.D. parallel to incident beam, R.D. vertical; (c) R.D. parallel to incident beam, N. vertical

1. The normal (N.) to the plane of rolling is parallel to the incident X-ray beam, while the rolling direction (R.D.) is kept vertical.
2. The transverse direction (T.D.) is parallel to the incident X-ray beam, while the rolling direction (R.D.) is kept vertical.
3. The rolling direction (R.D.) is parallel to the incident beam, while the normal (N.) is kept vertical.

Fig. 8.7 shows the relationship between the three different specimen orientations. The multitude of grains present in the rolled sheet diffract the X-ray beam in a similar manner to the randomly orientated pulverised crystals, except that darker arcs corresponding to the preferred orientation of the grains are superimposed. Fig. 8.7 also shows the stereographic projections of the Debye–Scherrer rings of an (*hkl*) lattice plane for each of the three specimen orientations.

In Fig. 8.8, three diffraction patterns corresponding to the specimen orientations of Fig. 8.7 are reproduced: these photographs are of a 0·005 in thick tough pitch copper sheet specimen annealed for 10 min at 200°C. The rolling schedule of the specimen is not known. Copper $K\alpha$ radiation at 35 kV and an exposure time of 30 min were used. In order to plot the pole figure, the powder rings must first be indexed. Since copper crystallises in a face centred cubic lattice, it follows that the innermost ring is a 111 reflection and the outer is a 200 reflection. Next the angles between the darker arcs and a vertical fiducial mark are measured, as shown in Fig. 8.7. For greater accuracy, the angular displacement of all four segments should be measured and the averaged readings tabulated. The results obtained from Fig. 8.8 are shown in Table 8.5.

Table 8.5.

Measured Data for the Texture Determination of a Copper Strip

Ring	Indices of reflection	θ	$90° - \theta$	N parallel to incident rays, R.D. vertical		T.D. parallel to incident rays, R.D. vertical		R.D. parallel to incident rays, N vertical	
				δ_1	δ_2	δ_1	δ_2	δ_1	δ_2
Inner	111	21·5°	68·5°	43°	63°	70°	90°	$\begin{cases}21° \\ 70°\end{cases}$	$\begin{cases}43° \\ 90°\end{cases}$
Outer	200	25°	65°	44·5°	62°	53°	70°	31°	61°

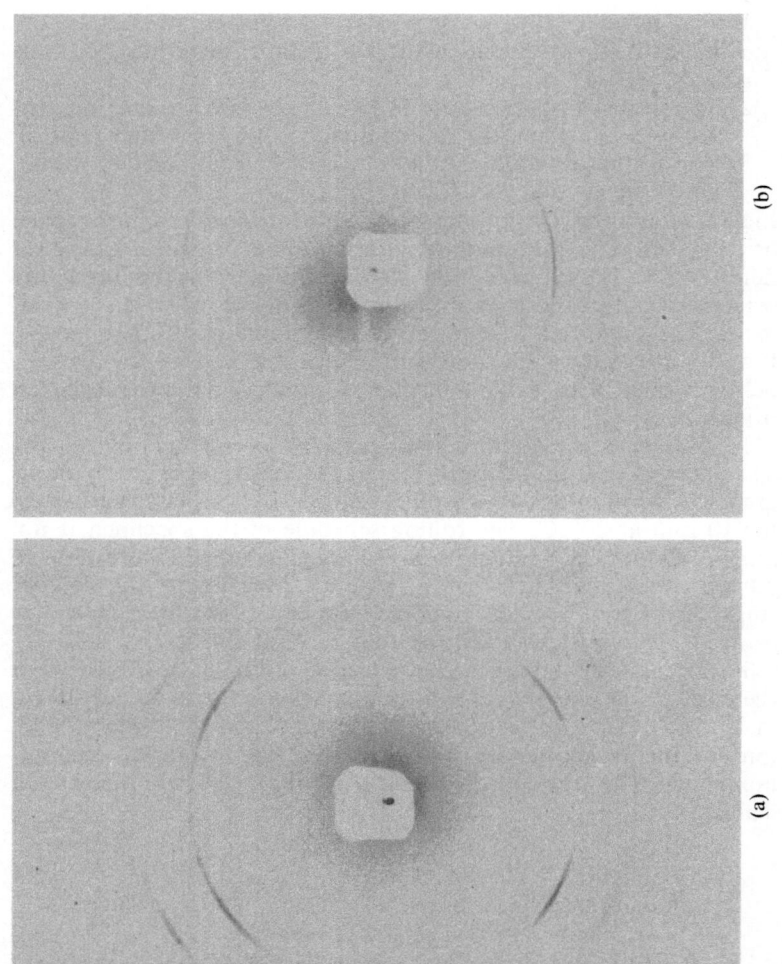

(a)

(b)

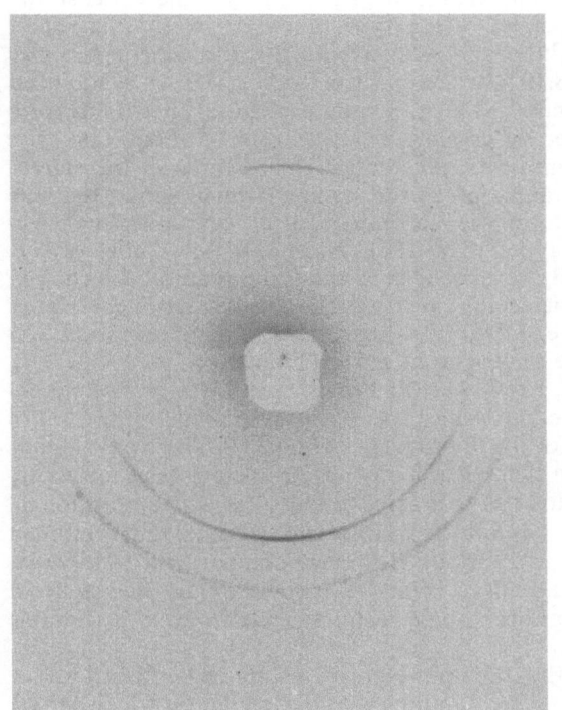

(c)

Fig. 8.8. Diffraction photographs of annealed tough pitch copper sheet at various orientations: (a) N. parallel to incident beam, R.D. vertical; (b) T.D. parallel to incident beam, R.D. vertical; (c) R.D. parallel to incident beam, N. vertical

From the tabulated measurements, the stereographic projections of the powder rings are drawn, as indicated in Fig. 8.9 and Fig. 8.10. Next, the darkened arcs are plotted from the known angular measurements. The shaded area, which is largely conjectural, shows the concentration of the {111} type poles in the stereogram of Fig. 8.9; similarly the probable distribution of the {200} type poles is shown in Fig. 8.10. By comparison of Fig. 8.9 to a standard stereogram of the cubic system, it can be deduced that the texture is mainly composed of crystals in a (100)/[001] orientation. The only deviation seems to be the central shaded area of the stereogram. Detailed examination of the same specimen shows that the main 'cube' orientation is accompanied by an annealing twin component of the type (122)/[212]. The {200} pole figure of Fig. 8.10 is not so readily recognisable as the stereogram of the cube in the (100)/[001] orientation, because the twin component rather distorts the picture. Obviously, the amount of information obtainable from three photographs is very limited.

For comparison, Fig. 8.11 and Fig. 8.12 show a pair of stereograms of annealed copper sheet which has been similarly treated to the specimen used for the photographs of Fig. 8.8. These pole figures were compiled with the aid of X-ray spectrometers designed to detect minute variations in diffracted intensities. The resolution of high-intensity peaks is incomparably clearer than that obtained by the photographic method. However, comparison of the two methods shows that, for a rapid and approximate detection of sheet texture, the photographic technique yields quite acceptable results.

The accuracy of the three-photograph technique described above is badly affected by the fact that the area of stereogram which can be scanned is limited. As an alternative, the specimen can be rotated around a suitable axis to allow an increased number of photographs to be taken, usually at 5° or 10° intervals. J. F. Custers has developed a simple and accurate photographic method, along these lines, for the determination of pole figures: the specimen is set up with the rolling direction of the sheet vertical, and the transverse direction is adjusted to lie an angle θ to the incident X-ray beam, where θ is the Bragg angle. The specimen is then rotated around its normal direction in steps of 5° or 10°. The experimental arrangement is shown diagrammatically in Fig. 8.13. This method offers the advantage of a constant absorption condition between successive photographs; this is achieved because the path length of the X-ray beam within the specimen is kept constant. The technique is very

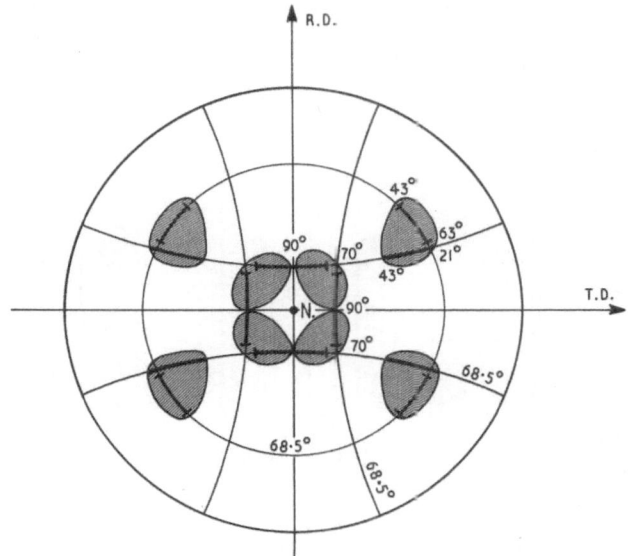

Fig. 8.9. {111} pole figure plotted from the data of Fig. 8.8

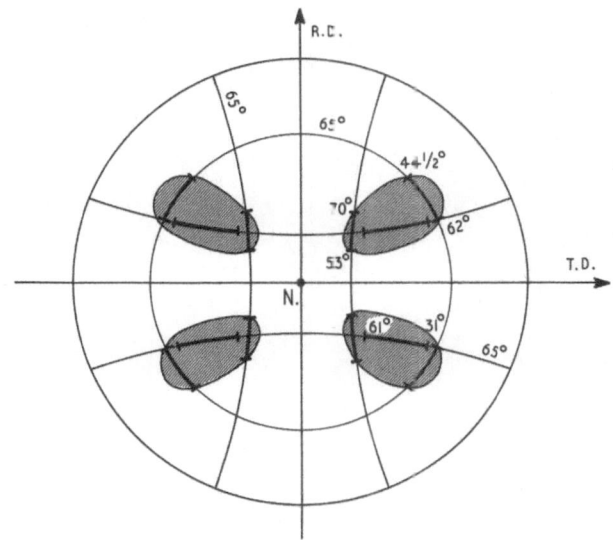

Fig. 8.10. {200} pole figure plotted from the data of Fig. 8.8

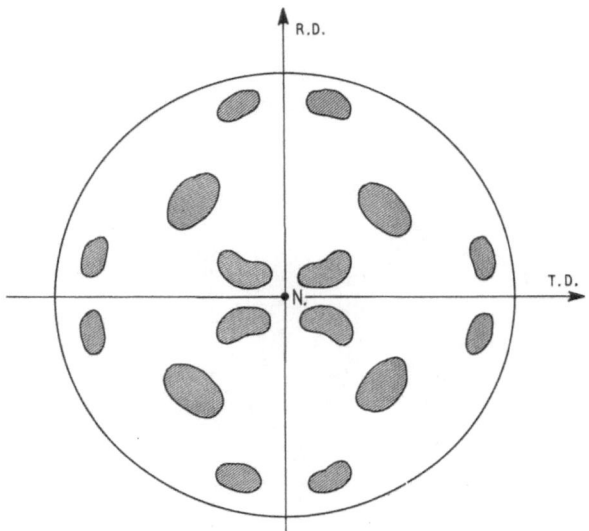

Fig. 8.11. {111} *pole figure of annealed copper sheet.* [*After* BECK *and* HU, J. Metals, **4,** *83 (1958)*]

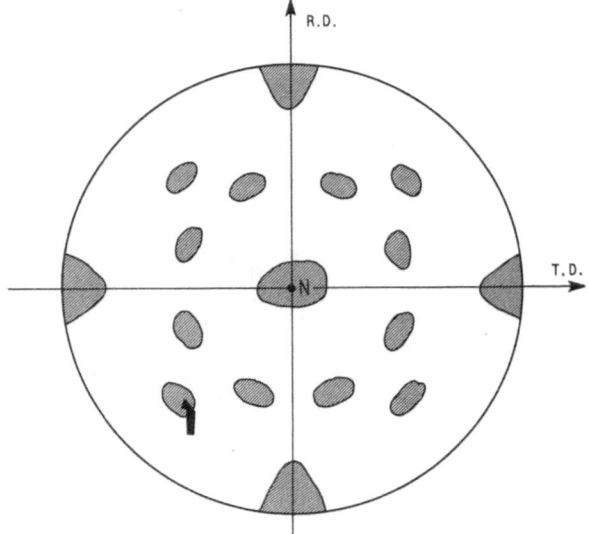

Fig. 8.12. {200} *pole figure of annealed copper sheet.* [*After* BECK *and* HU, J. Metals, **4,** *83 (1958)*]

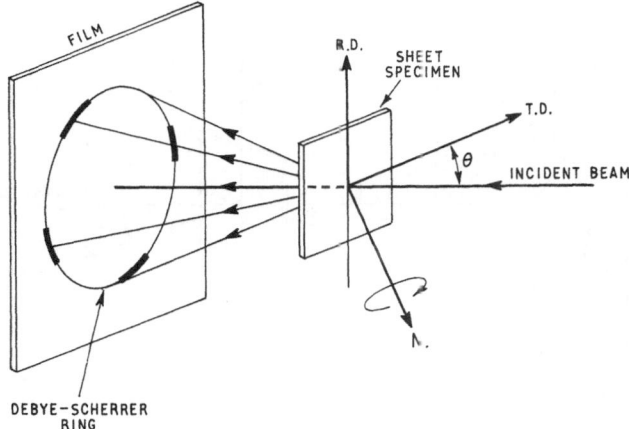

Fig. 8.13. Arrangement of a specimen for the Custers' method

suitable for highly absorbent specimens, since part of the Debye–
Scherrer rings are obtained by back-reflection instead of wholly
by transmission.

The pole figure is derived from the photographs by a similar
construction to that described for the three-photograph technique;
the difference is that, instead of only two pairs of small circles,

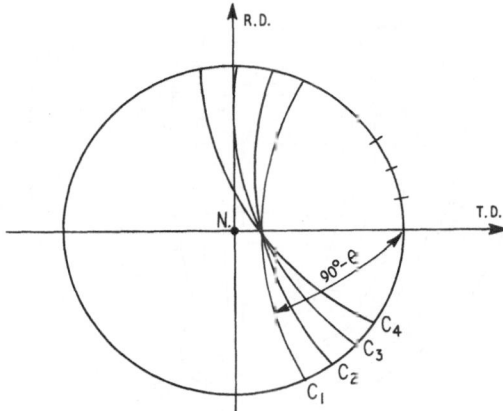

*Fig. 8.14. Stereographic projection of powder rings
obtained by the Custers' method*

173

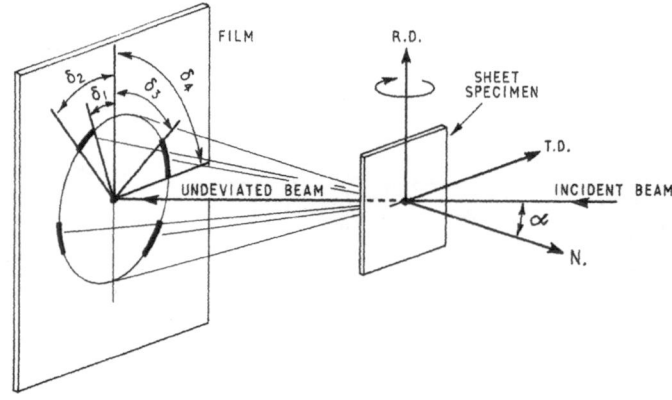

Fig. 8.15. Arrangement of a sheet specimen rotated through an angle α around the R.D. axis

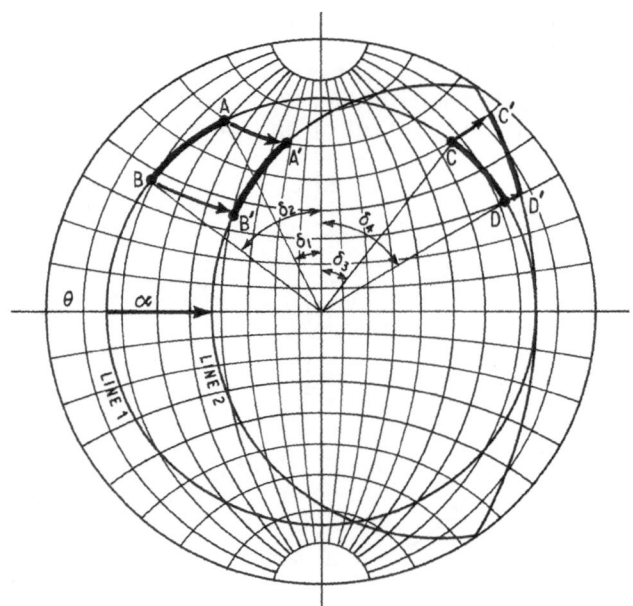

Fig. 8.16. Method of plotting a pole figure for oblique incidence

there are now as many pairs as there are photographs. The method of plotting is illustrated in Fig. 8.14. The stereographic plot of the Debye–Scherrer ring will be a small circle C_1. The other small circles correspond to the angular rotation of the specimen around the axis N. perpendicular to the rolling plane. The intense segments are then plotted on the small circles, as they were in Fig. 8.7. Fig. 8.14 shows that the rotating specimen sweeps out the whole area of the stereogram, with the exception of a small area of diameter 2θ in the centre. If this central area of the completed stereogram is required in detail, a 2θ rotation around the transverse direction will provide the necessary information.

When the specimen is rotated around axes other than the normal N. axis, the plotting of the stereogram becomes slightly more elaborate. In Fig. 8.15 the specimen has been rotated through an angle α around the R.D. axis. Suppose that the diffraction pattern is as shown in the diagram. In all probability, the positions and the lengths of the high-intensity arcs will not be symmetrical about the vertical axis owing to the oblique incidence of the beam. Fig. 8.16 illustrates the steps that should be followed in drawing a pole figure. In this hypothetical example, α is assumed to be 30°. When the specimen is arranged so that its normal axis is parallel to the incident X-ray beam, the stereographic projection of the Debye–Scherrer ring is concentric with the primitive circle and $\theta°$ inside it (line 1 in Fig. 8.16). When the specimen is rotated by the assumed 30°, the reflection circle rotates around the R.D. axis in the same sense as the specimen. To rotate the reflection circle, the points must be moved along the latitude lines of the Wulff net to the 30° position (line 2 in Fig. 8.16). (Note that, where the angle α exceeds the value of θ, the corresponding part of the displaced reflection circle falls 'behind' the reference sphere.) The intense segments AB and CD are then plotted on the reflection circle in the normal position (line 1), and are rotated by the requisite 30° along the latitude lines into the positions $A'B'$ and $C'D'$.

Since the texture of sheets is such that they are at least symmetrical about the R.D. direction, and often about the T.D. direction also, the $A'B'$ and $C'D'$ arcs may be reflected in the planes of symmetry to help fill in the pole figure. For this reason, the arc $C'D'$ is plotted on the 'front' hemisphere, instead of on the far side. Care should be taken in using these short cuts: if evidence is found for the absence of the symmetries described, then the short cuts must not be applied. If α is chosen in steps of 5° or 10°, the whole pole figure may be plotted with little or no guesswork.

175

It is nearly always advisable to try to plot the density changes, together with the dense areas, in a pole figure. Four different grades of darkening can usually be distinguished easily on diffraction photographs with the aid of a photodensitometer. This is simply a device for measuring the various degrees of blackening within the intense arcs of a powder ring. The method of plotting the pole figure is identical to the one illustrated above, except that the arc AB is divided into four sections of varying densities. The resulting contour areas are cross-hatched, to indicate the levels of pole density. This is in fact one of the best generally available methods of showing sharpness of texture without the use of expensive specialised equipment such as diffractometers.

In general, the photographic examination of preferred orientation does, of course, present manifold difficulties, the main one being the various absorption levels which result from the change in path length of the X-ray beam within the specimen as the setting angle α is changed. This factor inevitably introduces errors in estimates of the pole density. A simple and ingenious method, first proposed by Bakarian, of eliminating absorption variations is to grind the sheet specimen into a cylinder, thereby keeping the path length constant whatever the angle of setting may be.

A second serious difficulty occasionally arises when the grains are so large that they give no continuous powder rings and the intense segments blend too well into the spotty background. In order to eliminate this effect, more grains must be brought into reflecting positions – either by oscillating the specimen a few degrees, or by moving the specimen parallel to the film between successive exposures.

In the past years, many attempts have been made to construct special cameras which will move the specimen and the film together, so that one film will contain sufficient information for the whole of the pole figure to be plotted. Such cameras are not available on a commercial scale, for their design is too specific to be useful in the average X-ray metallurgical laboratory. The X-ray diffraction analysis provides one of the most powerful tools for the control and possible improvement of sheet metal textures. Faulty rolling schedules or annealing treatments may readily be detected by X-rays, improvements can be sought and the results again checked by diffraction techniques. Apart from metallurgical microscopes, no other methods provide the same flexibility and industrial usefulness as the technique of X-ray diffraction.

Bibliography

CLASSICAL AND X-RAY CRYSTALLOGRAPHY

BRAGG, W. H. and W. L., *The Crystalline State*, Vol. I-III, Bell, London (1933–1953)
LONSDALE, K., *Crystals and X-rays*, Bell, London (1948)
PHILLIPS, F. C., *An Introduction to Crystallography*, Longmans, Green, London (1963)
International Tables for X-ray Crystallography, The Kynoch Press, Birmingham (1952)

EXPERIMENTAL TECHNIQUES

AZAROFF, L. V. and BUERGER, M. J., *The Powder Method in X-ray Crystallography*, McGraw-Hill, New York (1958)
BUERGER, M. J., *X-ray Crystallography*, Wiley, New York (1942)
BUNN, C. W., *Chemical Crystallography*, Oxford University Press, London (1945)
CLARK, G. L., *Applied X-rays*, McGraw-Hill, New York (1955)
D'EYE, R. W. M. and WAIT, E., *X-ray Powder Photography*, Butterworth, London (1960)
GUINIER, A., *X-ray Crystallographic Technology*, Hilger and Watts, London (1952)
HENRY, N. F. M., LIPSON, H. and WOOSTER, W. A., *The Interpretation of X-ray Diffraction Photographs*, Macmillan, London (1961)
KLUG, H. P. and ALEXANDER, L. E., *X-ray Diffraction Procedures*, Wiley, New York (1954)

METALLURGICAL TECHNIQUES

BARRETT, C. A., *The Structure of Metals*, McGraw-Hill, New York (1952)
CULLITY, B. D., *Elements of X-ray Diffraction*, Addison-Wesley, Reading, Mass. (1956)

177

GUINIER, A. and FOURNET, G., *Small Angle Scattering of X-rays*, Wiley, London (1955)

TAYLOR, A., *X-ray Metallography*, Wiley, London (1961)

UNDERWOOD, F. A., *Textures in Metal Sheets*, Macdonald, London (1961)

Index